T0271831

Biomedical and Environmental Sensing

RIVER PUBLISHERS SERIES IN INFORMATION SCIENCE AND TECHNOLOGY

Volume 5

Consulting Series Editor

Kwang-Cheng Chen
National Taiwan University
Taiwan

Information science and technology enables 21st century into an Internet and multimedia era. Multimedia means the theory and application of filtering, coding, estimating, analyzing, detecting and recognizing, synthesizing, classifying, recording, and reproducing signals by digital and/or analog devices or techniques, while the scope of "signal" includes audio, video, speech, image, musical, multimedia, data/content, geophysical, sonar/radar, bio/medical, sensation, etc. Networking suggests transportation of such multimedia contents among nodes in communication and/or computer networks, to facilitate the ultimate Internet. Theory, technologies, protocols and standards, applications/services, practice and implementation of wired/wireless networking are all within the scope of this series. We further extend the scope for 21st century life through the knowledge in robotics, machine learning, cognitive science, pattern recognition, quantum/biological/molecular computation and information processing, and applications to health and society advance.

- Communication/Computer Networking Technologies and Applications
- Queuing Theory, Optimization, Operation Research, Statistical Theory and Applications
- Multimedia/Speech/Video Processing, Theory and Applications of Signal Processing
- Computation and Information Processing, Machine Intelligence, Cognitive Science, and Decision

For a list of other books in this series, see final page.

Biomedical and Environmental Sensing

J. I. Agbinya

*University of Technology, Sydney, Australia/French
South African Technical Institute in Electronics
Pretoria, South Africa*

E. Biermann

*Assistant Director, French South African Technical Institute
in Electronics/Tshwane University of Technology
Pretoria, South Africa*

Y. Hamam

*Scientific Director, French South African
Technical Institute in Electronics Pretoria
South Africa/ESIEE Paris, France*

F. Rocaries

*Director, French South African Technical
Institute in Electronics
Pretoria/ESIEE Paris, France*

S. K. Lal

*University of Technology
Sydney, Australia*

River Publishers

Routledge
Taylor & Francis Group

LONDON AND NEW YORK

Published 2009 by River Publishers
River Publishers
Alsbjergvej 10, 9260 Gistrup, Denmark
www.riverpublishers.com

Distributed exclusively by Routledge
4 Park Square, Milton Park, Abingdon, Oxon OX14 4RN
605 Third Avenue, New York, NY 10017, USA

Biomedical and Environmental Sensing / by J. I. Agbinya, E. Biermann, Y. Hamam, F. Rocaries, S. K. Lal.

Routledge is an imprint of the Taylor & Francis Group, an informa business

ISBN 978-87-92329-28-8 (print)

While every effort is made to provide dependable information, the publisher, authors, and editors cannot be held responsible for any errors or omissions.

Dedication

Dedicated to Janice J Brett and Gordon Hamilton Brett

Acknowledgments

Many researchers worldwide have been involved in the review of the materials that form these book chapters. The initial papers were reviewed for BroadCom 2008 in very short formats and have been extensively extended in greater details to provide great details compared with their initial short five page limits. Hence the chapters have been reviewed again to ensure the quality of the new materials. We acknowledge the expert reviewers for their excellent work. It is impossible to mention all of them by name.

The following organizations also made significant contributions towards the staging of BroadCom 2008 which has led to the book. The French South African Technical Institute in Electronics (F'SATIE), the Meraka Institute, Pretoria South Africa, Tshwane University of Technology, Pretoria South Africa, University of Technology Sydney, Australia and the Technical Program Committee of BroadCom 2008 played the leading roles in the planning of the conference. The following people also played specific roles:

Dr Prins Nevhutalu, DVC (Research, Development & Innovation), Tshwane University of Technology, Pretoria, South Africa

Mr Melvin Kamote, CEO, Alcatel-Lucent, Midrand, Gauteng, South Africa

Ms Soloshni Naido (deceased), Meraka Institute, Pretoria South Africa

Ms Denise Umuhoza, Meraka Institute, Pretoria South Africa

Dr Ntsibane Ntlatlapa, Meraka Institute, Pretoria South Africa

Dr Anwar Vahed, Meraka Institute, Pretoria South Africa

Mr Gaullame Noel, French South African Technical Institute in Electronics (F'SATIE)

Dr Karim Djouani, French South African Technical Institute in Electronics (F'SATIE)

Mr Anish Kurien, Tshwane University of Technology, Pretoria, South Africa

Preface

Rapid growth in computing, software engineering and broadband access world wide have enabled realistic biomedical services and applications. These have combined to enhance the investigative and diagnostic aspects of biology and medicine. Biomedical computing opens new doors for better health care delivery through bio-condition and behaviour monitoring of patients. They have also helped to realise visualisation of complex human biological patterns. Biomedical computing can provide deeper insights and richer support to medical personnel to diagnose medical problems and proffer solutions much faster and more efficiently.

Objectives and Prerequisites

The book is intended as an introduction to biomedical monitoring, diagnostic medicine and environmental sensing for post graduate and undergraduate teaching. It is also a vital research text book in biomedical engineering, sensor networks and security and environmental monitoring. Clearly it is a text material which provides a strong foundation in these areas for for Master students, Ph.D. candidates, researchers from academia and industries, telecommunication equipment manufacturers, educators and key technocrats.

The goal in this book is mainly to assemble in one place the techniques and technologies which make medical computing a modern reality. The book focuses on

- biomedical decision support systems
- biomedical sensing
- medical computing and visualisation
- environmental sensing applications

Modern biomedical computing is rooted in a rich broad range of application areas. Imaging needs from microscopy to mammography have motivated and relied on advances in imaging science. Medical data storage and access systems benefit from the study of information retrieval. Algorithms and software development are of key importance in areas such as genome sequence analysis and acquisition, which also depend on techniques from statistics and artificial intelligence. The book assumes basic understanding and overview of health condition monitoring, sensors and their applications and therefore delves into the subjects from a graduate point of view. No assumptions are made in terms of the fundamental mathematics involved, but it is clearly obvious that it will be better understood at and above the final year undergraduate degree level.

Organisation of Chapters

The text consists of fifteen chapters. Each chapter of the book can be read as a stand alone text on a specific sensing and biomedical technology. The first chapter presents and discusses a data driven decision support system. Decision analysis techniques are increasingly being applied to model and analyze dynamic decision problems in medicine. The chapter focuses on a multilevel approach for constructing data evidence based model for classification of different therapies and their effectiveness for clinical treatment. The authors present methods for aligning patient records, a mapping to clusters based on preprocessed sequences of critical events of patient treatment and algorithms for therapy planning support. Clusters of patient records medication profiles can be derived based on their methods. A method for therapy decision support is based on profiles to determine a sequence of ordered steps of possible medication for a given patient state.

Chapter 2 extends the thread of the discussions in the book by simplifying the complexities of modern anesthesia procedures. Methods used by a prototype (Diagnesia) of a decision support system for anesthetists are presented. During surgery, Diagnesia shows how to use patient data recorded to continuously estimate likelihood and unlikelihood of diagnoses, applying arguments for and against the different diagnoses, and presents the most probable diagnoses to the anesthetist. These provide decision support to the specialists whose sole responsibility is to ensure the good health of a patient undergoing surgery.

This facilitates faster decision making. Also by presenting information about the patient on a single interface at a given level of abstraction, together with some suggested inferences, the specialist can use this information, coupled with his/her expert knowledge to diagnose a situation.

Chapter 3 describes a non-invasive hyperthermia therapy device based on radio frequency techniques. Every year, the burden of benign and malignant tumours continues to increase including also the need for better and more effective cancer treatment modalities. Thermal ablation therapy is a minimally invasive technique in cancer management. The authors investigated the use of radiofrequency (RF) energy in cancer management, developed and tested a minimally invasive thermal probe that will effectively destroy volumes of pathological tumours by means of hyperthermia. The chapter highlights some of the key aspects of their work. The different strategies employed in thermal ablation therapy and the basic principles of RF ablation are briefly discussed. Next, the reaction of tissue to thermal injury at different temperatures is examined. Using experimental tissue samples from soft tissue, kidney, liver, lung and brain tissue, the results of tests carried out with the RF thermal probe to evaluate for effective tissue destruction capability are evaluated to determine the best power setting for effective tissue ablation over specified test period.

Chapter 4 addresses a new area and discusses methods for measuring brain functions. The measurement of the electrical signals on the scalp, arising from the synchronous firing of the neurons in response to a stimulus, known as electroencephalography (EEG), opened up new possibilities in studying brain functions in normal subjects. However it was the advent of the functional imaging modalities of positron emission tomography (PET), single photon emission computed tomography (SPECT), functional magnetic resonance imaging (fMRI), and magnetoencephalography (MEG) that led to a new era in the study of brain functions. An introduction to the Metabolism and Blood Flow in the Brain is given. This is followed by a more detailed explanation of functional MRI and how such experiments are performed.

Chapter 5 presents automatic processing of chromosomes. When chromosomes are segmented from image as objects, some of the chromosomes that are close to each other (touching chromosomes) or overlapped are detected as one object. Objects that have more than one chromosome, are processed separately from those have one chromosome. One stage of automatic processing of chromosomes is automatic separation of multi-chromosome images from one-chromosome images. In the chapter a solution using feed forward neural networks is discussed and explained.

Chapters 6 to nine discuss key techniques for human health condition monitoring. In Chapter 6 measurement of body surface potential is explained. Body surface potential mapping (BSPM) is a non-invasive functional imaging method for reconstructing electrophysiological information about the surface of the heart. BSPM provides more diagnostic information than the 12-lead ECG. De-noising body surface potential (BSP) signals leads to more accurate BSPM patterns therefore better diagnosis for heart problems. In the chapter, for de-noising of body surface potential signals FFT and also different types of wavelets are used as a de-noising tool. For the better performance evaluation of different de-noising tools a quantitative value called Signal Error Ratio (SER) is described. After de-noising all of the BSP signals, changes in BSPM patterns are also presented to show the effects of our method in de-noising BSPs.

Chapter 7 describes the development of wireless electroencephalogram (EEG) to be used in a fatigue countermeasure device for train drivers. EEG can be used as an indicator of fatigue. It is known that slow wave brain activities, delta (0–4 Hz) and theta (4–8 Hz), increase as an individual becomes fatigued, while the fast brain activities, alpha (8–13 Hz) and beta (13–35 Hz), decrease. This chapter proposes the design of a single channel wireless EEG device that is suitable for fatigue detection in train drivers. The EEG measurement was designed for bipolar montage recording. The recording in bipolar montage is obtained from the difference between two adjacent active electrodes that are adjacent to each other.

Chapter 8 focuses on remote monitoring of ECG based on the application of mobile terminals for monitoring patient conditions in a non-hospital setting. It describes the architecture of an economic wireless transmission system and an implementation of an effective algorithm, adapted to the mobile terminal, allowing a doctor to obtain the results of analysis of ECG data wirelessly.

Chapter 9 is an examination of a hybrid system for detecting driver drowsiness using piezofilm movement sensors integrated into car seats, seat belt and steering wheel. Statistical associations between increase in the driver drowsiness and the non-invasive and conventional physiological indicators are described. Statistically significant associations are explained for the analysed physiological indicators — car seat movement magnitude and (electroencephalogram) EEG alpha band power percentage.

Chapter 10 is organized to include a review of fundamental properties of wireless sensor networks, system components and the areas of biomedical applications of WSN applications. Biomedical applications of wireless sensor networks (WSN) are categorized into three areas:

- Patient habitat monitoring systems
- Patient monitoring systems (monitoring of biological systems)
- Actuator-enabled devices that affect a functioning system

The chapter overviews the privacy and security obstacles associated with biomedical applications including

- Limited resources
- Fault tolerance, interference, and attacks
- Confidentiality and
- Physical security

Chapter 11 is a full blown discussion of cryptographic methods for studying the security and key establishment schemes for distributed sensor networks (DSNs). The approach is first to provide an overview of research directions that constitute the security in DSNs. These include attacks, cryptography, key management, secure routing, secure applications, and intrusion detection. Then the chapter discusses in more detail about how the key agreement problem has been tackled via the investigation of key establishment and pre-distribution schemes. Finally, a proposed new pairwise key establishment scheme for clustered DSNs is given. The scheme is shown to have desired security properties and to be highly efficient, cost effective, and practical for resource-constrained sensor nodes.

Chapters 12 and thirteen address different aspects of planning of sensor networks. Chapter 12 addresses the problems of where and how best to place the sensors for optimum tracking of mobile objects. It also addresses the issues of sensor selection for best coverage in addition to optimising the coverage area. It intentionally takes a pragmatic approach and assumes that it is possible through good engineering to optimally place sensors to achieve an objective. Sensors do not radiate omni-directionally as is assumed by many authors on sensor networks. In fact for example PIR sensors are highly directive and have limited fields of views. Unplanned tracking wireless sensor networks for surveillance focus generally on ability to detect motion but miss the essential features of best coverage and range. By optimally placing the sensors, choosing appropriate sensors with the required field of view and with power and frequency planning the capabilities of very short coverage range sensors can be maximised and the scalability of such networks becomes significantly easier. Addressing of sensors in such networks is also overviewed and a new method for assigning unique addresses to the sensors based on arithmetic series is described in details.

Chapter 13 is related to Chapter 12 and covers scheduling and redeployment mechanisms in wireless sensor networks for both atomic event and composite event detection. An important issue in WSNs is energy management. Sensor nodes are battery powered and in general, they cannot be recharged. It takes a limited time before they deplete their energy and become nonfunctional. One of the major components that consume energy is the radio. A radio can be in one of the following modes: transmit, receive, idle, and sleep. A radio is in idle mode when the sensor is not transmitting or receiving data, and usually the power consumption is as high as in the receive mode. A radio is in sleep mode when both the transmitter and the receiver are turned off. Sensors may be equipped with different numbers and types of sensing components. They can detect an atomic event independently or they can cooperate to detect a composite event. Sensors can be put to sleep to save energy. In this chapter scheduling mechanisms that select a set of active sensors to perform sensing and data relaying while all other sensors go to sleep to save energy are discussed. Sensors in the active set change over time in order to prolong network lifetime. In addition redeployment mechanisms which exploit sensor mobility to relocate sensors to improve the initial deployment are given.

Chapters 14 and fifteen provide a rich overview of the mathematics associated with sensor-based stochastic processes and an application based on the methodologies. Chapter 14 gives a survey of the combination of classifiers and the basic principles of machine learning and the problem of classifier construction. It reviews several approaches to generate different classifiers as well as established methods to combine different classifiers. Then it introduces a novel approach to assess the appropriateness of different classifiers based on their characteristics for each test point individually.

Chapter 15 introduces an approach for modeling sensor-based environmental monitoring with an example, a structural deterioration of components of bridges for maintenance optimization purposes. The Markov chain model is used to drive the current bridge maintenance optimization systems. While this model results into solvable programming problems and provides a solution, there are a number of criticisms associated with it. The chapter highlights the shortfalls of the Markov model for bridge lifetime assessment and promotes the use of stochastic processes. In the Markov model, the condition of a bridge element takes discrete states and the transitions from one state to the other are modeled with a Markov chain. The chapter discusses the fundamentals and explains the applicability of the gamma process and other stochastic processes for modeling structural deterioration.

Contents

1

Data Driven Therapy Decision Support System

Witold Jacak and Karin Proell

Upper Austria University of Applied Sciences, Austria

Therapy modeling and planning are important components for optimal and cost-effective patient care. Therapeutic response of the individual patient not only relies on the selection of effective drugs but is also heavily influenced by appropriate drug dosage. A variety of intelligent techniques have been initiated to support physicians in deciding an optimal treatment for an individual patient.

In recent years, decision analysis techniques are increasingly being applied to model and analyze dynamic decision problems in medicine [1, 2]. Dynamic decision analysis and modeling frameworks are based on structural and semantically extensions of conventional decision models e.g. decision trees and influence diagrams, with the mathematical definitions of finite state Markov Stochastic Processes [3, 4]. Most approaches use Markov Decision Processes (MDP) to describe and solve decision problems in which the optimal choice has to be revised periodically in accordance with the evolution of the patient's conditions [1, 3, 4]. Unfortunately the adoption of MDP to model complex systems as medical decision problems is hampered by the difficulty in knowledge elicitation from a specific domain. In particular the traditional formulation of a MDP [4, 5] through its transition matrix imposes to specify a great number of parameters, whose meaning is not always understood promptly, and it does not allow us to represent explicitly the structured knowledge underlying the model.

Therefore a clinician relies on his knowledge of fundamental physiologic and pathologic processes to develop diagnostic methods and procedures and to investigate the effects of drugs and new treatments on real clinical cases. The clinician must have a precise representation of the clinical state of the patient and of relevant physiologic processes ongoing in the patient's body. Representations of those cognitive structures are based on clinical observations, case records and available empirical data.

A further commonly used formalism for knowledge representation is a *semantic network* with a *graph-grammar* approach to manage the complex graph transformations driven by information entries (patient data) and medical problem solving (Such a system is applied in Children's University Hospital Mainz, Germany, for therapy planning in pediatric oncology) [6]. Similarly to MPD approaches the complexity of semantic network systems grows with the increase of available patient data.

This chapter focuses on a multilevel approach for constructing a data evidence based model for classification of different therapies and their effectiveness for clinical treatment. We present methods for aligning patient records, a mapping to clusters based on preprocessed sequences of critical events of patient treatment and algorithms for therapy planning support. Having clusters of patient records medication profiles can be derived. A method for therapy decision support is based on such profiles to determine a sequence of ordered steps of possible medication for a given patient state. The decision system is described in Section 1.3.

Therapy planning can also be seen as a learning problem. In [7, 8] an extension to the Q-Learning algorithm is used to incorporate existing clinical expertise into the process of acquiring an appropriate administration strategy of rHuEPO to patients with anemia.

In order to synthesize the Q-Learning agent we need a patient state generalization function to provide generalized patient states as categories of patient observation vectors, which is presented in Section 1.2. For classification of medications we introduce a medication generalization function based on similarity classes of medications and a similarity function between two drug dosages. Both generalization functions are used for generalizing patient trials and the life long quality function. In Section 1.4 the Q-Learning Agent is presented.

In the first phase, a sequence of events called *patient trial* will be extracted from computer patient records (CPR). These events describe only one flow of therapy of a concrete disease. Each event is represented as a pair *(state, time)*.

The *state* does not only contain standard numeric parameters but can be extended with images (MR, RT, or photo) and text based linguistic descriptions. Based on such *state* we introduce the measure between states of different patients. We assume that each patient's state is represented in the global state space. Based on the measure the system calculates the best alignment between different patient trials. The alignment measure (score) calculates the distance between two sequences of patient states, which represents the similarity of flow of therapy [14].

This procedure is applied to each pair of patient trials stored in the Hospital Information Systems (HIS) concerning similar diseases. Based on the value of similarity a semantic network is constructed and divided into full-connected partitions. Each of these partitions represents a class of similar therapy and can be used for computer-aided decision-making. The clustering can be extended by integrating biomedical information such as gene expression data of microarray date for those patient sets.

1.1 Patient Record and Patient State Generalization

Patient Record Data based Patient State Distance

On the patient level of knowledge base the data from patient records should be preprocessed to obtain a compact representation of course of disease. Normally we have different sources of medical information concerning a couple of numerical data representing laboratory test results, RT images or linguistic text describing diagnosis and therapy. The general patient record can be represented as *critical event set* described by the *observation protocols set*

$$Patient\ Record = \{Critical\ Event_j | j = 1, \ldots, n\} \qquad (1.1)$$

The information in form of *critical events* is the notification that assists in ensuring compliance with practice guidelines and medication protocols. The clinical documentation contains a great number of medical parameters from which structured knowledge about patient state must be derived.

The clinician must have a precise representation of the clinical state of the patient and of the relevant physiologic processes ongoing in the patient. Each state represents a variable obtained from the factorization of the parameter space. Different methods for factorization (based on fuzzy sets) can be found in [9].

Each *critical event e* is represented as a pair *(state, time)*. The state does not only contain standard numeric laboratory test parameters v but also text based linguistic description of observations (observation protocol).

$$Critical\ Event = e = (state,\ time) = ((v,\ Observation\ Protocol),\ \tau)\quad (1.2)$$

where v is vector of test parameters, and τ is the time-interval between initial moment t_0 and current event's moment t, $(\tau = t - t_0)$.

Representation of such a cognitive structure as general patient state is based on clinical observations, case records and available empirical data. The classic method of state construction is the structured interview. Such interviews contain clinical findings and disease states. The interview results are used to get a formal observation protocol. The diagnostic checklists — structured interview — provide an automated method for assuring documentation of key symptoms and behavioral issues. Critical events allow clinicians to document the reason for deviating from standard treatment [10, 11, 12, 13, 15]. In psychopharmacology, for example, there are few procedures and laboratory tests that definitively establish a psychiatric observation. The International Classification of Diseases (ICD) and DMS provide standard criteria on which to base the observation protocol. The role of theoretical representations has been verified experimentally by several studies [12, 13].

In pediatric oncology for example, protocols typically cover time periods including initial diagnostics and assignment to different risk groups and therapy branches, intensive inpatient chemo- and radiotherapy phases for tumor remission, consolidation therapy, and outpatient long-term therapy to avoid tumor reoccurrence [6]. In this domain the observation protocol represents time-tables of chemo- and radiotherapy, long-term sequences and stratifications of therapy phases, and temporal constraints concerning the duration of therapy-intensive and therapy-free intervals.

In the first phase the patient record should be transformed into a sequence of critical events e representing the time course of a patient healing process called patient trials.

$$trial = (m, (e_1, \ldots, e_n))$$
$$= (medication, (state_1, time_1), \ldots, (state_n, time_n))\quad (1.3)$$

Each change of treatment or medications m establishes the new patient's trial of disease. Critical events describe only one flow of treatment for a concrete treatment of disease and medication. In order to compare different trials it is necessary to introduce a formal description of states to allow the calculation of distance or similarity between two states. It is obvious that a vector of numerical data $(v_j | j = 1, \ldots, n)$ is easy to compare. To make it possible to measure the similarity between images it is sometimes useful to map images to 3D models of the human body which can be used to obtain additional numerical data $(w_j | j = 1, \ldots, k)$ derived on various image contents. Such methods can be useful in treatment of dermatological diseases. Based on 3D models we cannot only calculate various geometrical parameters as for example field, contour length or shape but we also can automatically perform a classification of infected anatomical regions and a coding of disease (Figure 1.1). Additionally we assume that diagnosis will be transformed into standard code for example ICD-10.

After these preprocessing steps we can represent the state as a vector of numerical parameters and codes (i.e. diagnosis code *diag_code*).

$$e^i = (state^i, time^i)$$
$$= (((v_j^i | j = 1, \ldots, n), (w_j^i | j = 1, \ldots, k), diag\_code^i), time^i) \quad (1.4)$$

Based on such *state* we introduce the measure of distance $\rho: S \times S \to R^+$ between states of different patient (see Figure 1.2).

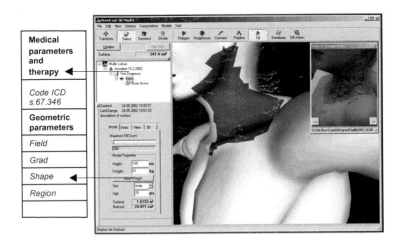

Fig. 1.1 Case-dependent patient state containing numerical data and 3D-model data.

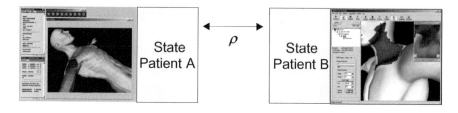

Fig. 1.2 The local measure between states of different patient.

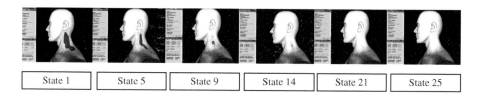

| State 1 | State 5 | State 9 | State 14 | State 21 | State 25 |

Fig. 1.3 Example of a patient trial.

We assume that each patient's state is represented in this global state space S. This distance represents the similarity of two states form different trials of the course of disease. An example of a patient trial for a dermatological disease is presented in Figure 1.3.

The images above are extracted from a concrete patient trial showing the healing process of a dermatological case in the neck region. Each state is based on a 3D model of the affected part of the body and a set of numeric and coding parameters representing results of assessments or treatment protocols. The 3D Model and the parameters are used to build the state vector at each point of time, which are used to calculate the measure between states of treatment in different patient trials.

Each change of treatment or medications establishes the new patient's trial of disease T. It means that there exists the function

$$M: \textit{Trials} \rightarrow \textit{Medications} \tag{1.5}$$

such that $M\,(trial) = m$. The patient trials are shown in Figure 1.4.

The distance function $\rho: S \times S \rightarrow R^+$ between states of different patient can be used to analyze similarity of patient state and for computing the distance function between different trials.

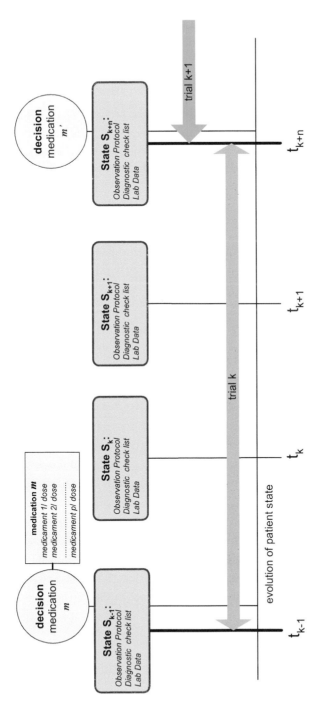

Fig. 1.4 Patient trials.

Other method for generalization of different patient states uses the self-mapping neural network to construct the *similar (generalized) state* from different patient states [16].

Patient State Generalization with Self-Mapping Neural Network

One of the main components of the therapy planning system is the module for building a function that gives generalized states from clinical patient records like patient data, clinical guidelines, outcome laboratory measures etc. Representation of such a cognitive structure as *generalized patient state* is based on clinical observations, case records and available empirical data.

We establish the generalized state function

$$\textbf{\textit{gen}}_{State} : \textit{Patient States} \rightarrow \textit{Generalized States Space} \qquad (1.6)$$

The only information available is the k-dimensional vector v of test parameters and factorized observations protocols $v = (v, \textit{Observation Protocol})$. Our goal is thus to construct a state generalization function to provide generalized patient states as categories of patient observation vectors. This leads to clusters of observation vectors with vectors in the same cluster being mapped to the same generalized patient state by the function $\textbf{\textit{gen}}$. The function $\textbf{\textit{gen}}_{State}$ that produces generalized states from observation vectors can be implemented as a neural network, so that new states can be incorporated to the implementation of $\textbf{\textit{gen}}_{State}$ by an unsupervised learning process [16]. The problem of obtaining generalized states from patient data to preserve proximity information among the observation vectors can be solved by a applying a neural network [17, 18, 19, 20] that forms clusters in its input space and produces a good representative of these clusters as output.

The network operates in combining the Kohonen clustering algorithm and the class creation and pruning methods incorporated in the fuzzy-ART and fuzzy-ARTMAP algorithms. The topology of the network consists of an input layer and an output layer with full connection between these two layers. The input layer has dimensionality k, and the output layer grows and shrinks as new category neurons (each representing a conceptual state) are added and deleted.

When an input vector is presented to the network, all the output neurons calculate their distance (i.e., the distance of their weight vector to the input) in parallel. The neuron with the smallest distance is the winner neuron.

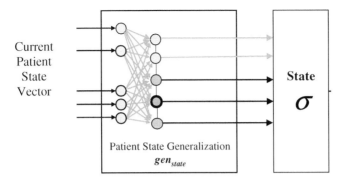

Fig. 1.5 Neural Network for patient state generalization.

At this point, we use an idea from the fuzzy-ART algorithm and check whether the winning neuron is close enough to be able to represent the input vector, or whether the input vector is so dissimilar to the winning neuron's weights (and thus to all the other categories as well) that it has to be placed into a new category. For this, we define a *similarity radius* as the maximum range an observation vector can be distant from the winning neuron's weights to be still considered close enough to fall into that category.

Category neuron training: If the observation vector v is within the similarity radius of the winning neuron's weights w than the weights of the winning neuron and the neurons in its neighbourhood are adjusted to reflect the new entry in this category by moving them into the direction of v.

Category neuron creation: If the observation vector is not within the similarity radius of the winning neuron, it is not similar enough to be included in one of the current categories, then a new category has to be created.

In this algorithm, this corresponds to the creation of a new output neuron that is fully connected to the input layer, and whose weight vector is equal to the sensor vector. The network starts with one output neuron that is created upon presentation of the first input pattern. Output neurons are created when an observation vector is found not to be within the similarity radius of any of the output neurons. It is clear that the smaller the similarity radius is, the more output neurons will be created because the criterion for similarity is stronger the smaller this radius is.

Category neuron pruning: To prevent a proliferation of output neurons, we include a pruning step that cuts output neurons. The pruning can be done according to two different criteria:

- a neuron is cut when it has not been the winning neuron for a given period of time, or
- a neuron is cut when it encodes too small a fraction of input vectors compared with the output neuron that encodes the most observation vectors.

The output associated with the clustering algorithm is not the output of the network, as these are only the similarity numbers for the various categories. Instead, the output of the algorithm is the weight vector of the winning output neuron. In other words, the output of the algorithm is the representative of the category that is most similar to the input presented to the network. If the observation vector is within the similarity radius of one of the output neurons, the output is the weight vector of this neuron; otherwise, it is the observation vector itself (which is also being incorporated in the network as a new category neuron).

The network presented above produces conceptual states from observation data. These generalized states are then used as inputs for the Q-Learning approach that models the effects of medication in generalized states space. The neural network that implements this function is presented in Section 4.

1.2 Therapy Planning Based on Patient Trials

Trial Level: Alignment of Trials

To store a patient's data and course of treatment, an electronic patient record as a source for retrospective inspection and medical decision-making is provided. There are many different approaches to store and generalize raw patient data. The generalization process depends on the kind of representation of knowledge for the medical decision-support system. One of most used formalism for knowledge representation is a *semantic network* with a *graph-grammar* approach to manage the complex graph transformations driven by the information entry (patient data) and medical problem solving [6]. For such knowledge representation it is necessary to use a similar representation of patient trials. The patient trial is constructed as a temporal subnet for the representation of the course of treatment. This temporal net represents medical events such as the administration of a drug through labeled, attributed nodes, while temporal relationships are modeled with labeled edges [6]. New patient data is represented as sub-graph and special *graph production rules* of the patient graph-grammar are used for mapping the new sub-graph into semantic patient graph. This production maps an inconsistency between two (diagnostic) data items, (such as a new laboratory value being not consistent to the current diagnosis) to the patient graph. After having found the matching region in the patient graph, the production generates a new node and connects it to the existing nodes via inconsistency edges and tests the inconsistency related data. [6]. The complex network transformation processes driven by knowledge acquisition, problem solving and data entry are controlled by a graph-grammar.

Semantic graphs of individual patients are not efficient for the analysis of similarities between different patient trials. We present here one of other methods, which can be used to analyze patient trials.

The most basic analysis task of trials is to ask if two trials can be related. There are many positions at which two corresponding patient states (entries) can be compared. However, in the case when the one of the trial has extra entries, then in a couple of places gaps have to be inserted to the second trial to maintain an alignment across such regions. When we compare sequences of patient states, the basic processes are considered as substitution, which changes entries in the sequence, and insertion or deletion, which adds or removes entries. Insertions and deletions are referred to as gaps. An example for state alignment can be seen in Figure 1.6. Based on the distance ρ between two generalized patient states (see Section 1.2), the system calculates the alignment distance (score) ρ^*, which describes the similarity (distance) between different patient trials.

$$\rho^* : S^* \times S^* \to R^+ \qquad (1.7)$$

where S^* is the set of ordered sequences of the states from the patient state space S. Each sequence $x = (x_1, \ldots, x_n) \in S^*$ has a respective sequence of events describing the concrete patient trial $(e_1, \ldots, e_n)$.

The measure $\rho^*(x, y)$ calculates the distance between these two sequences, which describes the similarity of the flow of the therapy. We apply a special sequence-matching algorithm, well known in Bioinformatics for biological sequence alignment [21].

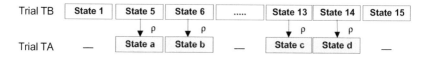

$\rho^* =$ minimal total score $=$ optimal alignment

Fig. 1.6 Alignment between two patient trials.

x_{i-k}	x	x	x_{i-1}	x_i
y_l	g	g	g	g

Fig. 1.7 Gaps Insertion.

The global measurement (score) we assign to an alignment is the sum of terms for each aligned (similar) pair of states (entries), plus terms for gaps (see Figure 1.7). We will consider a pair of trials (patient states sequences) x and y of lengths n and m, respectively.

Let x_i be the i-th state in x and y_j be the j-th symbol of y. Given a pair of trials, we want to assign a score to the alignment that gives a measure of the relative likelihood that the trials are related as opposed to being unrelated. For each two states x_i and y_j we can use ρ as measure of similarity of these residues. Let $n > m$ then we should add gaps g in the second trial to find the best alignment.

Gap penalties

We expect to penalize gaps. Each state x_i in the sequence x has an additional parameter τ_i which represents the time interval between the state x_{i-1} and x_i. The first state x_1 has $\tau_1 = 0$. For finding the penalty value of a gap at the i-th position of the trial y we use the knowledge about time intervals associated with each state. Let the last ungapped substitution with state y_l be on $(i - k)$-th position in the trial x.

The standard cost associated with a gap is given by

$$\rho(x_i, g) = K \exp(-(|\tau_i^x - \tau_l^y|)) = pen(x_i, y_l) \tag{1.8}$$

where τ_i^x is the time interval associated with state x_i and τ_l^y is the time interval associated with state y_l from trial y which was aligned with the state x_{i-k} from the trial x. The long insertions and deletions with different intervals of time are penalized less as those where the intervals of the time is quite the same.

Alignment algorithm

The problem we consider is that of obtaining an optimal alignment between two patient trials, allowing gaps. We can use the well known dynamic programming algorithm, which has many applications in biological sequences analysis [21]. The idea is to build up an optimal alignment using previous solutions for optimal alignments of smaller subsequences. The problem can be defined as follows:

> *Find the best alignment between sequences x^* and y^* (x^*, y^* represent sequences x and y extended by necessary gaps) such that global score*
>
> $$\rho^*(x, y) = \Sigma(\rho(x_i^*, y_i^*)) = min \qquad (1.9)$$
>
> *where $x_i^* = x_i$ or gap g and if x_i is aligned to y_l and x_{i+k} is aligned to y_u then $l < u$.*

We can use the known dynamic programming algorithm, which has many applications in the biological sequences analysis [21]. The idea is to build up an optimal alignment using previous solutions for optimal alignments of smaller subsequences. We construct a matrix F indexed by i and j, one index for each trial, where value $F(i, j)$ is the score of the best alignment between the initial segment $x_1, \ldots, x_i$ of x up to x_i and the initial segment $y_1, \ldots, y_j$ of y up to y_j. We can build $F(i, j)$ recursively. For details see [22]. Let us assume that we are only interested in matches scoring ρ less than some threshold T, it means that similarity between two states of the patients is very high. $F(i, 0)$ is the minimal (best) sum of completed match scores to the subsequence $x_1, \ldots, x_i$ assuming that x_i is in an unmatched region.

It is obvious that we expect that one trial contains the other or that they overlap. It means that we want a match to start on the top or left border of the matrix and finish on the right or bottom border. The initialization equations are that $F(i, 0) = 0$ for $i = 1, \ldots, n$ and $F(0, j) = 0$ for $j = 1, \ldots, m$. Now

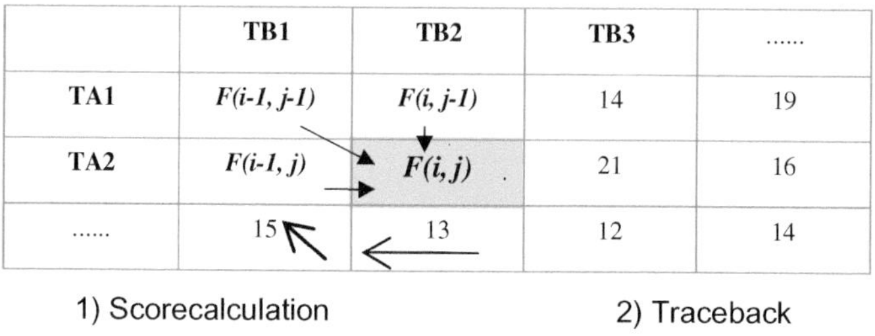

	TB1	TB2	TB3	......
TA1	*F(i-1, j-1)*	*F(i, j-1)*	14	19
TA2	*F(i-1, j)*	*F(i,j)*	21	16
......	15	13	12	14

1) Scorecalculation   2) Traceback

Fig. 1.8 Matrix for alignment calculation.

we calculate recursively the matrix value as

$$F(i,0) = min \begin{cases} F(i-1,0) \\ F(i-1,m) + T \end{cases}$$ (1.10)

$$F(i,j) = min \begin{cases} F(i-1,j-1) + \rho(x_i, y_j) \\ F(i-1,j) + pen(x_{i-1}, y_j) \\ F(i,j-1) + pen(x_i, y_{j-1}) \end{cases}$$ (1.11)

The calculation steps are presented in Figure 3.3. Let F_{min} be the minimal value on the right border (i, m) for $i = 1, \ldots, n$, and the bottom border (n, j) $j = 1, \ldots, m$. This minimal score is the measure of the similarity between the complete two trials x and y. i.e.

$$\rho^*(x, y) = F_{min}$$ (1.12)

To find the alignment itself we must find the path of choices that led to the minimal value. The procedure for doing this is known as a trace back. The trace back starts from the minimal point and continues until the top or left edge is reached.

This procedure is applied to each pair of patient trials.

Clustering of the Trial Space

Based on ρ^* we can build the distance matrix between each trial and use standard clustering methods [14] to obtain the similarity classes (cluster) of trials. The similarity cluster c is defined as follows:

- $c \subset Set\ of\ Trials$
- $(\forall x, y \in c)(\rho(x, y) < \varepsilon) \ldots$ (1.13)
- $card\ (c) \rightarrow max$

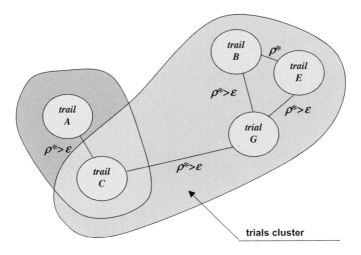

Fig. 1.9 Trials clustering.

where ε is the threshold value discriminating the similarity of two trials. For construction of similarity classes we build the graph which nodes are trials and its arcs is the relation ρ^*. The threshold value cuts different arcs. The similarity class is the maximal sub-graph of the graph for which its nodes are full connected (see Figure 1.9).

Profiling of Patient Trials Families

In the previous sections we have already created a set of trials belonging to a particular cluster. Such a cluster is called a trial family. Trials in a family are diverged from each other in the primary sequence of the course of a disease. For each cluster c the profile trial in form of *pair generalized initial state* and *generalized final state*

$$profile(c) = (s_{initial}^c, s_{final}^c) \qquad (1.14)$$

is established.

The *generalized initial/final state* is generated as most probably *factorized values vector* of laboratory parameters from initial/final states and text mining created *common context* of observation protocols from initial/final states of each trial from cluster c respectively.

Additionally, for each trial *trial* from cluster c the medications m i.e. treatment of drug dosages, therapeutic serum levels and indications are assigned in a unique way. The different medications m can be applied to obtain the generalized final state from generalized initial state.

Let $\mathbf{M}(c) = \{M(trial) \mid trial \in c\}$ be the set of medications applied in trials from cluster c. For each medication m_i from $\mathbf{M}(c)$ we can calculate the application probability as

$$p(m_i) = \mu(m_i)/\Sigma\{\mu(m_j) \mid m_j \in \mathbf{M}(c)\} \qquad (1.15)$$

where $\mu(m_i)$ is the frequency of medication m_i application in cluster c. The profile of cluster c is shown in Figure 1.10).

The distance measure ρ between the clinical states of patients may be used to introduce the new order relation between the clusters from cluster set C.

Let s_{fd} designate the state of a cured patient (destination state). We establish the preferable relation *pref* between the clusters

$$pref \subset C \times C$$

$$c_1 \, pref \, c_2 \Leftrightarrow \rho(s_{final}^{c1}; s_{fd}) + K\tau_{final}^{c1} < \rho(s_{final}^{c2}; s_{fd}) + K\tau_{final}^{c2} \qquad (1.16)$$

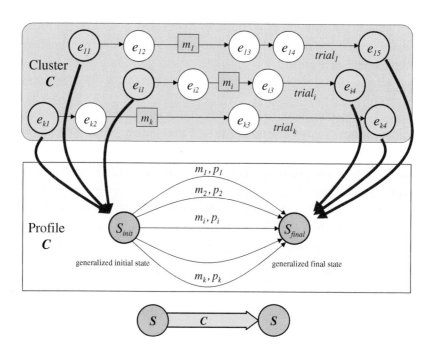

Fig. 1.10 Cluster and profile of cluster.

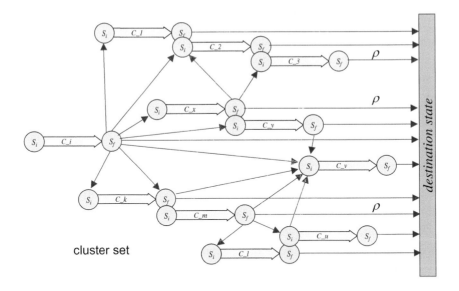

Fig. 1.11 Preferable relation between clusters.

where $(s_{init}^{c1}, s_{final}^{c1}) = profile\ (c_1)$, $(s_{init}^{c2}, s_{final}^{c2}) = profile\ (c_2)$ and $\tau_{final}^{c1}, \tau_{final}^{c2}$ are the time intervals associated with final events of both cluster respectively.

This relation *pref* establishes a linear order on the cluster set C. The ordered cluster set C is shown in Figure 1.11.

We assume that for each patient there exists generalized additional information such as age, gender, weight etc., and if available biomedical parameters. These static parameters are factorized and represented as fuzzy linguistic variables with appropriate membership functions. The set of fuzzy description of patient parameters can be clustered with help of fuzzy similarity relation [9].

As an additional support for diagnostics gene expression data can be used. Using micro-arrays, we can measure the expression levels of more genes simultaneously. These expression levels can be determined for samples taken at different time points during a biological process or for samples taken under different conditions. For each gene the arrangement of these measurements into a vector leads to what is generally called an *expression profile*. The *expression vectors* can be regarded as data points in high dimensional space. Cluster analysis in a collection of gene expression vectors aims at identifying subgroups - *clusters of co-expressed genes*, which have a higher probability of participating in the same pathway. The clusters can be used to validate or combine the cluster to prior medical knowledge. Many clustering methods are available, which can be divided into two groups: first and second generation algorithms. The first generation algorithms are represented by hierarchical clustering algorithms, K-means clustering algorithms or self-organizing maps. These algorithms are complicate in use and often require the predefinition of more parameters that are hard to estimate in biomedical praxis. Another problem is that first generation clustering algorithms often force every data to point into a cluster. It can lead to lack of co-expression with other genes. Recently new clustering algorithms have started to tackle some of limitations of earlier methods. To this generation of algorithms belong: Self-organizing tree algorithms, quality based clustering and model-based clustering [18, 20, 23].

Self-organizing tree algorithms combine both: self-organizing maps and hierarchical clustering. The gene expression vectors are sequentially presented to terminal nodes of a dynamic binary tree. The greatest advantage is that the number of clusters does not have to be known in advance. In quality based clustering clusters are produced that have a quality guarantee, which ensures that all members of a cluster should be co-expressed with all other members of these clusters. The quality guarantee itself is defined as a fixed threshold for a maximal distance between two points between clusters. Based on these methods it is possible to generate clusters on the gene expression states.

Let *CPR* be a gene expression cluster, which contains data obtained from micro-arrays, in the patient record set. Each micro-array is connected with one or more patient trials. The cluster *CPR* can be mapped to the patient trial space C. This results in two patterns, which can be used for classifying each trial. On the one side the pattern based on trial alignment - on the other side the pattern based on gene expression and statistic patient information can be used. Both patterns can be combined for creation fine classes containing very similar trials with high co-expression of genes and statistic patient information (see Figure 1.12) [20].

Based on these data we introduce the set of medications $\mathbf{M}(c)$ and set of patient information cluster $\mathbf{CPR}(c) = \varphi^{-1}(c)$ for each cluster c. Finally the therapy model can be represented as

$$T\_Model = (C, \{profile(c) \,|\, c \in C\}, \{\mathbf{M}(c) \,|\, c \in C\},$$
$$\{\mathbf{CPR}(c) \,|\, c \in C\}, pref) \tag{1.17}$$

The general structure of the therapy model is sketched in the Figure 1.13.

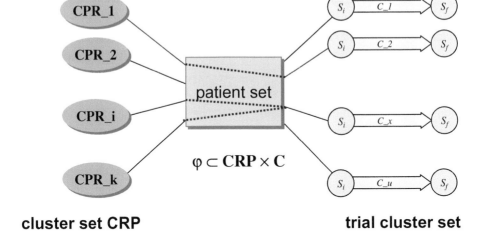

Fig. 1.12 Mapping CPR clusters and trial clusters.

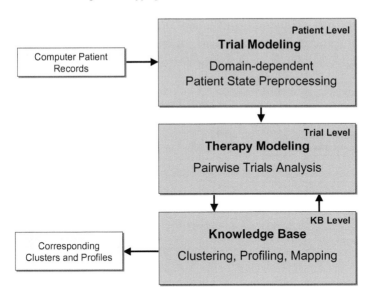

Fig. 1.13 Therapy model structure.

Therapy Planning Support

There are many different computer based support systems for diagnosis and therapy planning using different knowledge representation [1, 3, 6, 10, 11].

One of them is for example CEMS [10] or TheMPO: A Knowledge-Based System for Therapy Planning in Pediatric Oncology [6]. CEMS - Clinical Evaluation and Monitoring System - was developed to provide comprehensive support for clinical services in a psychiatric hospital. It represents a psychopharmacology monitoring system. The system provides decision support and automated monitoring for each key component of care i.e. assessment diagnosis and treatment and consist of four modules; treatment standards — pharmacotherapy guidelines, diagnosis checklists — DCL, and information alerts and outcome assessment.

Based on the first observation and CEMS procedures including laboratory tests we can establish the treatment and standard medication m_x for a new patient x. Additionally the clinical state of patient s_x and statistic patient information cpr_x can be established.

We can use our therapy model to evaluate the proposed treatment m_x or to find a more preferable treatment for this disease case.

Algorithm for evaluation and monitoring of clinical treatment: The evaluation procedure is realized in following constraints satisfaction steps (Figure 1.14). At first we look for the cluster in which the static data of patient

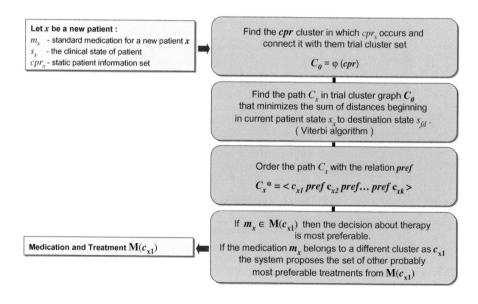

Fig. 1.14 Algorithm for evaluation and monitoring of clinical treatment.

x fits and we connect it to the trial cluster set. Based on this trial cluster we look for the path C_x in trial cluster graph that minimizes the sum of distances beginning in current patient state s_x to destination state s_{fd}. This path will be ordered with the relation *pref*. If current medication belongs to the first medication set in ordered path, then the decision about therapy is most preferable. If the current medication belongs to a different cluster the system proposes the set of other probably most preferable treatments.

This algorithm can be used to support the decision process for effective therapy planning in a clinical evaluation and monitoring system. The accuracy of the evaluation obviously depends on the number of trials used to construct the trial clusters. Model uncertainty about the trials sets in the clusters should be taken into account by adding posterior probabilities for the effectiveness of a therapy suggestion.

The concatenation of various trials in combination with extended computational methods could be used to gain deeper insight into the course of specific diseases depending on variations of medication dosages. The profile of cluster could be used to monitor a new patient's therapy course and produce alerts when deviations from the expected therapy path are observed.

1.3 Reinforcement Learning for Therapy Planning

Therapy planning for a specific disease is heavily affected by various levels of uncertainty concerning a patient's response to a therapeutic treatment. Underlying PD/PK models are often too complex to use in clinical practice, sometimes individual patients respond to drugs in an unexpected manner.

Therefore data driven approaches based on Reinforcement Learning like Q-learning strategies get more and more popular in planning individualized drug therapy. These strategies do not rely on PD/PK models for therapy planning but use clinical observations, case records and available empirical data for drug selection and dosage.

In [7, 8] the application of Reinforcement-Learning methods are used to incorporate existing clinical expertise into the process of acquiring an appropriate administration strategy of rHuEPO to patients with anemia. In [24] a reinforcement learning approach is used to extract optimal structured treatment interruption strategies directly from clinical data without using an accurate mathematical model of HIV infection dynamics. A framework for the application of Q-learning in therapy planning is presented in the next sections.

Generalization of medications and dosages

For each critical event e_i of patient trial (see Section 1.1), the medications m_i i.e. treatment of drug dosages, therapeutic serum levels and indications are assigned in a unique way. The medication can be represented as set $m_i = \{(c_{ij}, \alpha_{ij}) \mid j = 1, \ldots, k\}$ where c_{ij} is a chemical component of medicament and α_{ij} is its dosage.

For classification of medications it is necessary to introduce a similarity function between two medications m_k and m_j, s: $M \times M \rightarrow \mathfrak{R}$;

Let m_i' be a set of components of medication, $m_i' = \{c_{ij}, \mid j = 1, \ldots, k\}$ and $I_{ik} = m_k' \cap m_i'$ then the similarity function can be described as:

$$s(m_k, m_i) = 1 - \beta_{ik} \gamma_{ik} \tag{1.18}$$

where $\beta_{ik} = |I_{ik}| / max\{|m_i'|, |m_k'|\}$ and $\gamma_{ik} = min\{\Sigma \alpha_{cj}^i | c_{ij} \in I_{ik}, \Sigma \alpha_{cj}^i | c_{ij} \in I_{ik}\} / max\{\Sigma \alpha_{cj}^i | c_{ij} \in I_{ik}, \Sigma \alpha_{cj}^i | c_{ij} \in I_{ik}\}$.

It is easy to observe that $s(m_k, m_k) = 0$ and $s(m_k, m_i) = s(m_i, m_k)$. The similarity function realizes the tolerance relation on the set of M. This tolerance relation can be used to construct the similarity classes of medications as: The $M_\varepsilon \subset M$ is the tolerance class of medication if and only if

$$(\forall m_k, m_i) \in M_\varepsilon)(s(m_k, m_i) < \varepsilon) \tag{1.19}$$

The similarity classes — clusters of medication and similarity function are used to build the medication generalization function gen_{Med} (gen_{Med}: $M \rightarrow M_\varepsilon$) as

$$gen_{Med}(m) = \arg(\min_{M\varepsilon}(\max\{s(m, m_k) \mid m_k \in M_\varepsilon\})) \tag{1.20}$$

Q-Learning based Decision Support System for Therapy Planning

For each patient trial we can calculate a generalized trial as

$$\text{TRIAL} = ((gen_{Med}(m_i), gen_{State}(s_i)) \mid i = 1, \ldots, n)$$
$$= ((\mu_i, \sigma_i)) \mid i = 1, \ldots, n) \tag{1.21}$$

and $\delta \tau_i < \delta \tau_{i+1}$ and gen_{State}

The decision support system is based on Q-Learning [5] which is a popular learning method to select actions from delayed and sparse reward. The goal of Q-learning is to learn strategies for generating whole action sequences, which maximize an externally given reward function. The reward may be delayed and/or sparse, i.e. reward is only received upon reaching the goal of the task or upon total failure.

Let S be the set of all possible patient states and **gen**$_{State}$ be the generalization function mapping the current patient state s into the generalized state of patients. We use now the restrictive Markov assumption, i.e. we assume that at any discrete point of time, the system obtains the complete generalized state. Additionally, let $\mathbf{M}_\varepsilon$ (medication clusters) be the action set of the system. Based on the adequate state **gen**$_{State}(s)$ the system picks a generalized medication $\mu_\varepsilon \in \mathbf{M}_\varepsilon$, where $\mathbf{M}_\varepsilon$ is the set of medication clusters. As the result, the patient state changes. The trainer receives a scalar reward value, denoted by $r(\mathbf{gen}_{State}(s), \mu_\varepsilon)$, which measures the action performance. Such a reward can be exclusively received upon reaching a designated goal or upon total failure, respectively.

The Q-Learning algorithm should find an action strategy

$$\pi: S \to \mathbf{M}_\varepsilon \tag{1.22}$$

mapping from patient-states S to actions $\mathbf{M}_\varepsilon$, which, when applied to action selection, maximize the so called *cumulative discounted future reward*. For fast finding the best action in current state s the key of Q-Learning is to learn a value function Q for picking the actions:

$$Q: S \times \mathbf{M}_\varepsilon \to \Re \tag{1.23}$$

maps percept by agent conceptual states s and actions μ_ε to scalar utility values. In the ideal case $Q(\mathbf{gen}_{State}(s), \mu_\varepsilon)$ is, after learning, the maximum cumulative reward one can expect upon executing action μ_ε in state s. The function Q schedules actions according to their reward. The value $Q(\mathbf{gen}_{State}(s), \mu_\varepsilon)$ grows with the expected cumulative reward for applying action μ_ε in the current state s. The value function Q, after learning, allows generating optimal actions by picking the action which maximizes Q for current state s, i.e.

$$\pi(s) = \arg(\max\{Q(\mathbf{gen}_{State}(s), \mu_\varepsilon) \mid \mu_\varepsilon \in \mathbf{M}_\varepsilon\}) \tag{1.24}$$

The values of Q have to be learned over the whole lifetime of the system acting in the same disease. The function Q can be realized as the complex neural network. Initially, all values $Q(\sigma, \mu)$ are set to zero. During learning values are incrementally updated, using the following standard recursive procedure. Suppose the agent just executed a whole action sequence which, starting at some initial state σ_0 and led to a final state σ_F which reward $r(\sigma_F, \mu_F)$. For all steps i within this episode, $Q(\sigma_i, \mu_i)$ is updated through mixture of the values of subsequent state-action pairs, up to the final state. This standard procedure has the following form [26].

$$Q^{new}(\sigma_i, \mu_i) = \begin{cases} +\mathbf{N} \text{ if } \mu_i \quad \text{final action, positive result of treatment} \\ -\mathbf{N} \text{ if } \mu_i \quad \text{final action, negative result of treatment} \\ \psi(1-\lambda)(\max_\mu Q(\sigma_{i+1}, \mu)) + \lambda Q^{new}(\sigma_{i+1}, \mu_{i+1}) \end{cases}$$
$$(1.25)$$

where $\psi \leq 1$ is the discount factor.

Such Q-learning learns individual strategies independently, ignoring the opportunity for the transfer of knowledge across different sequences of medications. The architecture of the Q-Learning decision support system is presented in Figure 1.15.

Our approach considers various dosages of one specific drug and accounts for the fact that drug prescription can change during therapy by switching to medications with different active components.

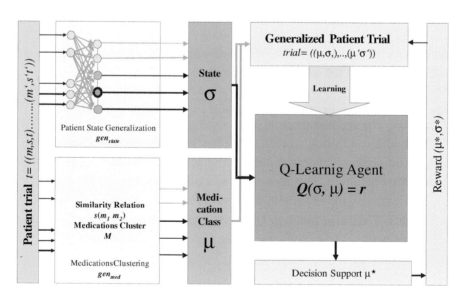

Fig. 1.15 Q-Learning based Decision Support System.

1.4 Final Remarks

Evidence-based Systems gain a more important role in decision support for medical practice. This chapter presents two approaches for a multilevel knowledge base system for therapy decision support combined with bioinformatics components. In the first level a sequence of events called patient trial is extracted from computer patient records. These events describe one flow of therapy for a concrete disease. Each event is represented by state and time. We introduce a measure between states, which is used to calculate the best alignment between different patient trials. The alignment measure calculates the distance between two sequences of patient states, which represents the similarity of the course of disease. On the second level based on similarity-value classes are introduced by using specific clustering methods. These classes can be treated more exactly by combining the information about gene expression data on microarrays and other patient data. This leads to finer clustering containing similar trials - called trial families.

The trial clusters can be used to support the decision process for effective therapy planning in a clinical evaluation and monitoring system. The accuracy of the evaluation obviously depends on the number of trials used to construct the trial clusters. Model uncertainty about the trials sets in the clusters should be taken into account by adding posterior probabilities for the effectiveness of a therapy suggestion. Furthermore the general patient data could be extended by adding information about the genotype of the patient such as relevant gene expression patterns observed in specific disease cases to obtain more diversification on cluster level. The second approach presented uses Reinforcement learning techniques (Q-Learning) for finding strategies for optimal drug selection and dosage.

References

[1] C. Cao, T. Leong, and A. Leong, "Dynamic Decision Analysis in Medicine: A Data Driven Approach". Int. Journal of Medical Informatics, 1999.

[2] W. Jacak and K. Proell, "Data Driven Therapy Modeling", Proceedings of the International Multiconference (I3M 2006), Barcelona, Spain, 2006.

[3] P. Magni, R. Bellazzi, and F. Locatelli, "Using Uncertainty Management Techniques in medical Therapy Planning: a Decision-Theoretic approach", Applications of Uncertainty Formalisms A. Hunter and S. Parsons, (Eds). LNCS, Springer, 1998.

[4] S. Miksch, K. Cheng, and B. Hayes-Roth, "An Intelligent Assistant For Patient Health Care", Proceedings of the First International Conference on Autonomous Agents (Agents'97), ACM Press, 1997.

[5] N. Lavrac, "Machine learning for data mining in medicine", Plenary invited talk at AIMDM'99, Aalborg, 20-24 June 1999. Proceedings of the Joint European Conference on Artificial Intelligence in Medicine and Medical Decision Making, pages 47–62, Springer Verlag, 1999.

[6] R. Müller, M. Sergl, U. Nauerth, D. Schoppe, K. Pommerening, and H.-M. Dittrich, "TheMPO: A Knowledge-Based System for Therapy Planning in Pediatric Oncology", Computers in Biology and Medicine, vol. 27(3), 177–200, 1997.

[7] A.E. Gaweda, M.K. Muezzinoglu, G.R. Aronoff, A.A Jacobs, J.M. Zurada, and M.E. Brier, "Incorporating prior knowledge into Q-learning for drug delivery individualization", Machine Learning and Applications, 2005.

[8] A.E. Gaweda, M.K. Muezzinoglu, G.R. Aronoff, A.A Jacobs, J.M. Zurada, and M.E. Brier, "Using clinical information in goal-oriented learning", Engineering in Medicine and Biology Magazine, IEEE, March-April 2007.

[9] H. Teodorescu, A. Kandel, and L. Jain, "Fuzzy and Neuro-Fuzzy Systems in Medicine", CRC Press, 1999.

[10] J. Bronzino, "Expert system in psychiatry: A review". Journal of Intel. Systems, 1993.

[11] R.A. Dunstan, R. Devenish, and R. Kevill, "A Computer-Guided Diagnosis System for Transfusion Medicine Laboratories", School of Biomedical Sciences and Academic Computing Services, Curtin University of Technology, 1997.

[12] C. Safran and G. Herrmann, "Computer based support for clinical decision-making", Med. Decision. Making, 1999.

[13] S. Smith and J. Park, "Therapy Planning as Constraints Satisfaction Problem: A Computer based Antiretroviral Therapy Advisor for the Management of HIV", Medinfo, 2000.

[14] Y. Moreau, F. De Smet, G. Thijs, K. Marchal, and B. De Moor, "Adaptive quality-based clustering of gene expression profiles", Bioinformatics, 18(5), 2002.

[15] F. Sonnenberg, C. Hagerty, and C. Kulikowski, "An architecture for knowledge based construction of decision models", Med. Decision Making, 1994.

[16] T. Kohonen, Self-organizing Maps, Springer-Verlag, 1997.

[17] S. Thrun and T. Mitchell, "Integrating inductive neural network learning and explanation-based learning", Proc. Of IJCAI'93, Chamberry, France, 1993.

[18] M.E. Brier, J.M. Zurada, and G.R. Aronoff, "Neural network predicted peak and trough gentamicin concentrations", Pharmaceutical Research 1995 Mar; pp. 406–412, 1995.

[19] J.D.M. Guerrero, E.S. Olivas, G. Camps Valls, A.J. Serrano Lopez, J.J. Perez Ruixo, and N.V. Jimenez Torres, "Use of neural networks for dosage individualisation of erythropoetin in patients with secondary anemia to chronic renal failure", Computers in Biology and Medicine, 2003.

[20] A.A. Jacobs, P. Lada, J.M. Zurada, M.E. Brier, and G.R. Aronoff, "Predictors of hematocrit in hemodialysis patients as determined by artificial neural networks", Journal of American Society of Nephrology 12, 2001.

[21] R. Durbin., S. Eddy, A. Krogh, and G. Mitchison, Biological sequence analysis, Cambridge University Press, 1998.

[22] R. Merkl and S. Waack, Bioinformatics Interactive, Wiley, 2001.

[23] V. Aris and M. Recce, "A method to improve detection of disease using selective expressed genes in microarray data", Methods of Microarray Data Analysis, Kluwer Academic, pp. 69–80, 2002.

[24] D. Ernst, G.-B. Stan, J. Gongalves, and L. Wehenkel, "Clinical data based optimal STI strategies for HIV: a reinforcement learning approach", Decision and Control, 2006 45th IEEE Conference on, Volume, Issue, 13–15 Dec. Page(s): 667–672, 2006.

2

Diagnesia: A Prototype of a Decision Support System for Anesthetists

John Kizito

Rijksuniversiteit Groningen, The Netherlands

2.1 Introduction

Patients who need to undergo surgery will need some type of anesthesia to go along with it in order not to feel conscious pain. The anesthetist will choose a certain type of anesthesia depending on a number factors. In some cases, you may require the anesthetic only at the region of operation (local anesthesia). Otherwise, sometimes procedures can not be performed unless the patient is completely anesthetized (general anesthesia). This chapter focuses on the case of general anesthesia.

When patients are anesthetized, they are unconscious and unaware of pain. They also do not feel conscious pain and are unable to move. They can not recall happenings during this period they have been anesthetized [1]. Because of this, patients undergoing surgery need to be monitored for the sake of their health and provide an environment for the surgeon performs his/her work.

Both anesthesia and surgery influence vital functions of the patient. Surgical effects include pain, blood loss, loss of water and electrolytes (for example by evaporation from the wound surface). Anesthetic effects include vasodilatation (dilation of a blood vessel, as by the action of a nerve or drug), decrease of myocardial contractility (the capability of the heart's muscular tissue to contract or cause contraction), suppression of autonomic nervous function, depression of spontaneous ventilatory drive, relaxation of respiratory muscles and so on. Anesthesia has an effect on the nervous system (which results in losing sensation) or the brain cells (which causes loss of consciousness).

Because of the consequences of all such effects, patients must be carefully monitored to ensure their survival and that they wake up after the anesthetic in good health.

The aim of this research therefore is to provide decision support to the specialists whose sole responsibility is to ensure the good health of a patient undergoing surgery. This facilitates faster decision making. By presenting information about the patient on a single interface at a given level of abstraction, together with some suggested inferences, the specialist can use this information, coupled with his/her expert knowledge to diagnose a situation.

2.2 The Theatre

There are a number of monitors used by the Anesthetist (person responsible for monitoring the patient — note that this is different from the person performing the surgery) to observe the state of the patient. Records read from these devices contain patient's physiological data collected during the surgery as well as registrations of certain events, medical history, and drug administration, and are recorded at certain intervals of time during the anesthetic. The anesthetist may monitor the patient by continuously observing some variables. Whereas on the one hand we have the surgeon's work, on the other hand, we have the Anesthetist's, which facilitates the latter.

The most commonly monitored variables include the patient's heart rate, blood pressure, respiratory rate, oxygen saturation, amplitude of plethysmogram (or amplitude pulse oximeter), end-tidal carbon dioxide, respiratory minute volume, tidal volume, respiratory pressures (inspiratory pressure/peak respiratory pressure and expiratory pressure), compliance, oxygen concentration of gas mixture, anesthetic concentration of gas mixture. Extensive monitoring may be required for some patients. This may be due to other factors like health condition and the type of surgery being carried out [1]. Such relevant information is displayed on monitoring devices and may be recorded at certain intervals of time.

2.2.1 The Anesthetist

As explained earlier, the anesthetist is a specialist who administers an anesthetic to a patient during treatment. There is a major distinction between an anesthetist and an anesthesiologist. Anesthesiologists are physicians specializing in anesthesiology. According to the Canadian Anesthesiologists' Society [2], the term *anesthetist* is used to refer to a person who administers an anesthetic. In some places, anesthetics are provided by only trained physicians. However, both anesthetists and anesthesiologists are highly trained and capable of administering anesthetics.

In [3], Bendixen [4] lists the following tasks for the anesthetist: "providing freedom of pain during surgery, record keeping, measurement and control of the vital functions, estimation of anesthetic depth, transfusion and fluid therapy". In The Netherlands anesthetics are always administered by a physician (anesthesiologist) and an anesthesia-trained assistant, who is not allowed to administer an anesthetic alone. In this chapter, the term *anesthetist* is used to refer to any personnel trained to administer anesthetics (anesthetists, anesthesiologists, and anesthesia-trained assistants). During surgery, it is the responsibility of the anesthetist to ensure the patient's well being and compensate for the effects of surgery and the anesthesia.

Without under-estimating the risks involved, anesthetic complications rarely occur and this can make the process of continuously observing the state of the patient boring which may result in low vigilance. Usually, the anesthetist expects everything to go on normally; however, this can be deceiving as complications can occur rapidly causing the anesthetist to make a decision in a very short period of time. Ballast [3] states that not all problems that occur during anesthesia are predictable. Unexpected problems do not occur frequently but when they occur, severe damage may be caused. Ballast gives examples such as: "allergic reactions, sudden blood loss, cardiac arrhythmias, breathing circuit disconnection, kinking of the endotracheal tube or accidental extubation". Together with other factors that relate to the wellbeing of the anesthetist, boredom and low vigilance can cause human error during the decision making process.

2.2.2 The Patient's State

The state of the patient is partially represented by the values displayed on monitoring devices. To define the patient's state, a number of variables are measured. Examples include: blood pressure, heart rate, and end-tidal carbon dioxide ($ETCO_2$). Normally, the anesthetist reads these values and uses his/her expert knowledge to diagnose situations. A variable whose value gets out of the normal/acceptable range could be an argument *for* or *against* a certain diagnosis. Not only do these variables tell the state of the patient but also other factors like preoperative status of the patient, physical appearance of the patient (e.g., sweating), previous actions taken on the patient (including treatment/drugs antecedently given), and so on. Such factors influence the state of the patient and should also be taken into account. The state of the patient mostly determines the decision-making behavior of the anesthetist.

Consequently, we divide the possible states of the patient in three categories: familiar to the anesthetist, urgent (i.e., life threatening) or requires diagnosing. In the Familiar state, the state is common to the anesthetist and he/she can opt for the typical treatment for this known diagnosis.

The urgent state is unfamiliar to the anesthetist and there is not enough time to investigate the cause of the problem since the situation is life threatening. Because the state of the patient is not familiar, no diagnosis is available. The anesthetist needs to give a treatment in order to take back the patient's state into a stable one, even without knowing the cause of the problem.

The third (diagnosing) state is similar to the urgent state. This state is also unfamiliar but not urgent. There are states of a patient that do not necessarily have a typical treatment — they cannot be diagnosed like in the familiar state. The state of the patient can be so complicated that the patient data being read gives no clear indication of a known diagnosis. This chapter will focus on the familiar state.

We have reduced the problem space to a finite number of diagnoses. In cases that are not urgent, there is time to process some information. Using the production rules built into the system, it will display a maximum of 5 most probable diagnoses with the ability to give the arguments *for* and/or *against* if needed. The anesthetist may use this abstract information, the patient data displayed on the monitoring devices, and his/her expert knowledge to make a diagnosis. Since there is time to make a diagnosis, communication with other members (or seeking for help) is facilitated by making information visible.

2.3 State of Art

In an earlier attempt to offer decision support in the theatre, methods that anesthetists use to diagnose problems that occur during surgery as well as a set of diagnoses that span most of the anesthesiological daily practice were investigated. A knowledge system prototype that continuously estimates likelihood and unlikelihood[1] of diagnoses in a set of 18 diagnoses, based on relative input parameters was then developed in Delphi 5 [5, 6]. The work presented in this chapter is an improvement of this prototype.

The developed system was tested with 12 chosen realistic situations that were also diagnosed by a panel of anesthesiologists. The diagnoses of the panel were used as a standard to compare the system's judgments with. In 11 test cases (92%), the knowledge system generated the same most probable diagnosis as the panel. In the 12th test case, the panel suggested two probable diagnoses, while the knowledge system only generated the second option. In 6 more test cases, the panel suggested 2 probable diagnoses. In three of these, the system did not recognize the second possibility, which was always a *low anesthesia* level. In the other three, the system suggested a general problem because it could not distinguish between two or more specific problems from the same category. In general, the system showed a high sensitivity of 92% and a very reasonable selectivity of 60% [6]. Sensitivity was calculated as the number of times a certain diagnosis was correctly indicated as the most probable and selectivity (or specificity) was estimated by the reciprocal of the average total number of diagnoses with an estimated likelihood higher than 20%.

Earlier attempts [6] succeeded in designing an algorithm for computing the likelihood (and unlikelihood) of diagnoses, which is the basis of our improvements made to the prototype. These methods will be presented in the next chapter. Some of the lacking areas were completeness of decision rules and usability issues. In this chapter, we do not present issues related to usability. A number of design decisions made during our discussions affected the working of the prototype and thus caused the modifications to the methods used by the prototype to present the information.

2.4 Methods

In this section we present the methods and models used by our decision support system, Diagnesia, in order to provide relevant information to the anesthetist.

---

[1] The phrase *unlikelihood of X*, as used in this chapter, refers to the chance or probability that X is not the probable diagnosis

2.4.1 Inputs and Outputs

When diagnosing situations, Anesthetists normally think in a certain manner. For example, if blood pressure were observed to be low, one would think: *Is it Hypovolemia?* But if the heart rate is also low, then: *May be not!* and so on. In this case, *low blood pressure* is an indicator for *Hypovolemia* while *low heart rate* is a counter indicator (for the same). Some arguments could be confirmatory as well. We thus have a set of variables that act as indicators for the different diagnoses. These indicators may belong to one of the following categories: cardio-vascular, ventilation, fluids, pulse-oximeter, and *others*.

But this is not enough; we still have a problem of defining what we mean by *low*, *normal*, or *high*! For instance one could say that 40–150 bpm is normal for a heart rate and 80–180 mmHg for the systolic blood pressure. However, these ranges can vary depending on other factors like patient's age and health. Children may have higher normal heart rates than old people; people who smoke might have lungs that do not perform as well as for non-smokers; sportspeople keep physically fit and may have different normal values from an average person who doesn't ensure physical fitness.

Despite these variations, anesthetists think in terms of: *the heart rate is low; the blood pressure is high but acceptable;* and so on. Consequently, we decide to categorize the values of these variables into 5 groups: low, low normal, normal, high normal, and high. A value may be *normal*, then increase to *high* or reduce to *low*. In case a value increases a bit but is still acceptable, we introduce an intermediate group, *high normal*. Correspondingly, we introduce *low normal* for values that get a bit low but still acceptable. At the extremes (*high* and *low*), all anesthetists should agree that the value is high or low and not acceptable. Thus the input to Diagnesia is the set of such indicator values and the output, for a given instance, is a set of most probable diagnoses, ranked as described in the next section.

2.4.2 Methods and Models

Using the indicators as inputs, Diagnesia builds rules that are used to continuously estimate likelihood and unlikelihood of diagnoses in the set by computing corresponding evidence probabilities. Every diagnosis has a set of indicators and counter indicators, each with a certain strength measured on

a 4-point scale (1 strongest, 4 weakest). The indicators are used to estimate
the evidence probabilities.

We define a rule as a combination of the indicator and other information concerning that indicator. Such information includes: the diagnoses supported, whether or not the indicator is *true*, weight of the indicator for each
of the supported diagnoses, and so forth. We then estimate the likelihood and
unlikelihood of diagnoses using the following steps:

   i. Initialize all rules by setting them to *false* and their corresponding
     evidence for all diagnoses to 0 (*zero*). The assumption here is that
     any indicator can be an indicator for any of the diagnoses in the
     set

  ii. For each rule and for all diagnoses, compute a score with which the
     rule can support the likelihood or unlikelihood of the correspond-
     ing diagnoses. This is achieved using the weights attached to the
     rules for the particular diagnoses. See Equation 1 and Equation 2.

Equation 1: $indices[diagnose] = \frac{weight}{|weight|} * 2^{(4-|weight|)}$

[2]**Equation 2:**

$$maxScore[diagnose, pro|con, 0 +/- indices[diagnose]]$$
$$:= maxScore[diagnose, pro|con, 0 +/- indices[diagnose]]$$
$$+/- indices[diagnose]$$

  iii. Set the rules that are correct to *true*. That is to say, if the heart rate
     is low, then we set the rule "*low heart rate*" to *true*.

  iv. For each of the diagnoses (n), use Equation 3 and Equation 4 to
     cumulatively compute the probability of the likelihood (or unlike-
     lihood). The general idea is to compute this probability as a ratio of
     the total score of all rules (expected to contribute to the evidence)
     that are true to that of all rules expected to make a contribution
     for the evidence in question.

---

[2]In equations, *pro* (and a plus sign) correspond to likelihood whereas *con* (and a minus sign) to unlikelihood

Equation 3:

$$p = gpMax[0 +/- indices[n]]^*d[n].pMax[pro\,|\,con]^*\,factor^*$$
$$\times \frac{indices[n]}{\max\ Score[n, pro\,|\,con, 0 \pm indices[n]]}$$

Equation 4: $d[n].kans[pro\,|\,con] := d[n].kans[pro\,|\,con] +/- p$

where gpMax (based on meaning attached to the weights), pMax (based on the diagnosis), and factor (based on the indicator value) are used to control the magnitude of the quotient.

2.4.3 Worked Example

This example is meant to give the reader a basic understanding of the methods presented in this section. Note that the meaning of the weights presented (1 strongest and 4 weakest) is ensured using a simple transformation to a geometric progression (GP) by Equation 1. This transformation can be seen more clearly in Table 2.1. Note also that the weights of counter indicators are negative and this sign is preserved by the expression $\frac{weight}{|weight|}$ in the formula (e.g., -1 maps to -8, and so on).

For our example, let us assume a scenario in which all indicators have normal values except that the compliance is low. Compliance is an indicator for pulmonary embolism, respiratory obstruction, pneumothorax, diffusion defect, backward, and muscle rigidity. We thus have the following rules fired (or being *true*):

  i. Low compliance
 ii. High saturation ($\equiv$ normal saturation. Counter indicator for backward failure, hypervolemia, severe hypervolemia, bad ventilation, and diffusion defect)

Table 2.1 Transformation of weights to a GP by Equation 1.

Weight	Index
1	8
2	4
3	2
4	1

iii. Normal heart rate (counter indicator for backward and forward failure)

iv. Normal blood pressure (counter indicator for backward failure, forward failure, and hypervolemia)

v. Normal heart rate and blood pressure (counter indicator for backward failure, forward failure, and hypervolemia)

vi. Normal MAC (counter indicator for deep anesthesia)

It follows that the rule *Low compliance* is fired with a factor equal to 1 and for this rule we have that:

$$indices[n] = \begin{cases} 1 & for\ n = backward\ failure,\ diffusion\ defect, \\ & and\ muscle\ rigidity \\ 2 & for\ pulmonary\ embolism,\ pneumothorax\ and \\ & respiratory\ obstruction \\ 0 & elsewhere. \end{cases}$$

According to Equation 2, we have the respective values in the maxScore array and the corresponding probabilities for the diagnoses in question as shown in Table 2.2. n represents any diagnosis. Only the 1st, 2nd, 4th and 8th positions of the maxScore array are presented since these are the only possible values of Equation 1 and therefore the only used locations of the array. Let us take a closer look at the computations of the probability of pulmonary embolism. In Table 2.3, we have the indicators for pulmonary embolism, their corresponding weights, and index values for each of the rules as computed by Equation 1. Since only one of these five rules (*Low compliance*), whose index value is 2, has been fired, the probability of pulmonary embolism is $0.5 * 1 * 1 * \frac{2}{4} = 0.25$ (Equation 3) as shown in Table 2.2. Correspondingly, if the compliance were to be *low normal*, Equation 3 would be used with a

Table 2.2 MaxScore and probability values for *Low compliance* scenario.

Diagnosis, n	maxScore [n, pro]	maxScore [n, con]	Likelihood probability	Unlikelihood probability
Pneumothorax	{2, 2, 0, 0}	{0, 0, 0, 0}	0.500	0.000
Respiratory obstruction	{2, 2, 0, 0}	{0, 0, 0, 0}	0.500	0.000
Muscle rigidity	{1, 0, 0, 0}	{0, 0, 0, 0}	0.250	0.000
Pulmonary embolism	{0, 4,12,0}	{0, 0, 0, 8}	0.250	0.000
Backward failure	{1, 0, 8, 0}	{2,10,8, 0}	0.250	0.383
Diffusion defect	{1, 0, 4, 0}	{0, 2, 0, 8}	0.250	1.000

Table 2.3 Pulmonary embolism probability computation for *Low compliance* scenario.

Indicator/Rule	ruleFired?	Weight	indices [n = *pulmonary_embolism*]
Low saturation	No	2	4
Forward failure	No	3	2
Low compliance	Yes	3	2
Low or falling ETCO$_2$	No	2	4
Increase in blood pressure	No	2	4
Total (maxScore[n, pro])			{0, 4, 12, 0}

factor of 1/2 making the probability 0.125. The corresponding probability of *hypersensitivity reaction* is $1 * 1 * 0.5 * \frac{4}{4} = 0.5$ due to *bronchospasm* that has a probability of $1 * 1 * 0.5 * \frac{4}{4} = 0.5$ caused by the probability of *respiration obstruction*.

In conclusion, it should be noted that the weights attached to the indicators and the meaning attached to the weights have very significant implication and thus a great contributor to the computation of the probabilities that are used to estimate the likelihood and unlikelihood of the diagnoses.

2.4.4   Key Design Decision

Diagnesia supports the anesthetist's decision making by suggesting a number of diagnoses. Perhaps only one of these is relevant for the anesthetist to diagnose the situation; however Diagnesia cannot be entirely selective, as it would then be assuming the decision-making. The challenge therefore that comes with support systems which deal with multiple disorders is the metric used to give the different options a ranking.

Szolovits and Pauker [7] describe a sequential approach of dealing with disorders that have overlapping findings. In this approach, all hypotheses are considered to be competitors. We then try to find the most probable diagnosis. When found, we proceed to find the second, third, and so forth. The hindrance with such an approach is that the program is unaware of the existence of more than one diagnosis at the beginning and thus findings that are not relevant to the primary disorder can easily confound the diagnostic process [8].

Szolovits *et al.* [8] state that "to deal with diseases whose findings overlap or interact, a program's best strategy is to use pathophysiologic reasoning that links diseases and findings through a network of causal relations." In this approach, the program builds a composite hypothesis that explains all the clinical findings. Competing hypotheses would be constructed in cases where several groups of disorders are consistent with the current clinical state. This approach however can not be adopted in a similar manner for single-disorder cases.

The approach taken in Diagnesia is a combination of these two. In Diagnesia, each fired rule suggests a hypothesis (or a number of hypotheses). By presenting each disorder with its level of likelihood, the system makes some selectivity to some extent even though this selectivity is entirely made by the anesthetist. On addition, more than one rule can be fired at the same time. Since each of these hypotheses claim to be of relevancy, we choose to find out the chance that one or more of them makes a correct claim. The 5 most probable diagnoses are thus determined by selecting the top most 5 diagnoses when arranged in descending order of the difference: *likelihood — unlikelihood*. For instance, if diagnosis *A* has likelihood of 1 and unlikelihood 0.5 and diagnosis *B* has likelihood 0.6 with unlikelihood 0, diagnosis *B* will have higher precedence than *A* in this list. Since the system presents a number of most likely disorders, the idea of considering a possibility of multiple disorders has also been adopted. Thus, one assumption to note here is that "the difference *likelihood — unlikelihood* is a good representation of the most probable diagnosis" and therefore provides relevant information to the anesthetist.

2.5 Results

In order to test our methods, an approach for testing the prototype was formulated. In this section, we present this approach and how/why the results expected can be a basis of an evaluation of the aspects of the system that we intend to test. The main intention in making the test here is to answer the question: *Does the system give a representation of the patient's state based on the patient's data for any given such state, which (representation) can be traced back to the data?*

2.5.1 Approach

It should be noted that the factors that influence the representation given by the system include the information used (input) to make calculations; the weights assigned to indicator; and the algorithm used to make the calculations.

We design a clinical picture (Figure 2.2) that can be used to simulate a given state of the patient and combines monitoring on one screen. This picture

Fig. 2.1 Anesthesia monitoring display.

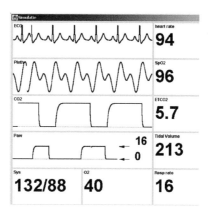

Fig. 2.2 Simulated anesthesia monitoring display (hypoventilation). Heart rate 94 bpm, saturation 96%, ETCO$_2$ 5.7 kPa, peak airway pressure 16 cmH$_2$O, positive end-expiratory pressure (PEEP) 0, systolic blood pressure 132 mmHg, diastolic blood pressure 88 mmHg, oxygen concentration 40%, tidal volume 213 ml, and respiratory rate 16.

is therefore not a real monitor screen but a simulated screen obtained by combining displays from different monitoring devices (say Figure 2.1). So, this clinical picture displays data in such a way that anesthetists are already familiar with. By use of such a simulated picture, we can easily alter the input data (which input data can also be fed into the prototype) and simulate cases of disorders for testing purposes.

Using such a picture, we can present a patient state, simulate the same state using Diagnesia, and test whether or not the state given by Diagnesia is a representation of the patient state on the clinical picture. This picture can also be used to compare the reaction of the anesthetists when presented with the picture and when presented with an output screen shot (perhaps of the same intended state) from Diagnesia.

2.5.2 Pre-test

In order to ensure that the clinical picture presented in 2.5.1 above is a representation of the actual monitoring devices (in other words, gives the same amount of information) and that anesthetists would have the same reaction when presented with the same picture, Ballast [3] conducted a test. In his test, he simulated 12 disorders and presented the corresponding pictures together with an unordered list of the diagnoses.

He then requested two very experienced anesthetists to match the diagnoses with the clinical pictures. One of the anesthetists matched all pictures correctly except for two diagnoses (tachycardia and sepsis). He assigned tachycardia to sepsis and vice versa. According to Ballast, the difference was quite clear and he supposed the anesthetist would very likely recognize his mistake. The same anesthetist, in hypoventilation, had expected a *low saturation* as an additional symptom. According to Ballast, this is possible but will not always be the case. The other anesthetist matched all pictures correctly except for ventilation failure and air embolism. He assigned ventilation failure to air embolism and vice versa. Again, Ballast claims the difference was quite clear and that a second look would surely convince the anesthetist.

Considering that both anesthetists mismatched 2 out of 12 cases, we can conclude that the clinical picture is 83% a presentation of the monitoring devices. We can also say that since all picture-diagnose combinations were correctly assigned by at least one of the anesthetists, and Ballast is convinced that the few misses were simple mistakes, the clinical picture correctly represents the monitoring devices.

2.5.3 Test Approach

In this sub-section, we discuss the approach to be used to test Diagnesia. The main goal is to test the prototype's representation of the patient's state by comparing the information from the state of art monitoring devices with that provided by the prototype. A clinical picture as shown in Figure 2 will be used to represent the data from monitoring devices and an output screen from Diagnesia when fed with the same patient data will be used for comparison.

A given state of a patient can be simulated by feeding the corresponding data into the simulator and obtaining a clinical picture. After feeding the same patient data into Diagnesia, we present the clinical picture and the output screen to a panel of anesthetists. Our intention is to check if the anesthetists will make the same diagnosis for both the clinical picture and the output screen. Note that the clinical picture makes no logical transformation on the data — it simply presents exactly what the monitoring devices display whereas Diagnesia transforms the patient data into information at a certain level of abstraction.

A number of disorders may be simulated, tested and the results obtained used to test whether or not the system gives a representation of a given state of the patient. Cases in which the specialists (anesthetists) reach different conclusions or totally disagree with the suggestion of the system should be investigated so that improvements (to the prototype) and/or further research can be made.

Apart from the pre-test results presented in 5.2, we have only discussed an approach to the main test in this section. It is however, highly recommended that the test be carried out and results obtained be used for further research and/or developments.

2.6   Discussion and Conclusion

In this chapter, we have presented methods and models used by a system to enhance the decision making process of the anesthetist by improving his/her situation awareness. It was noted that anesthetists are such skilled specialists that building a system to do their work or offer support with their work is not an easy task since we have to put into consideration the strategies they use in decision-making. It is evident that specialists have problems in making reliable probabilistic decisions and calculations [9]. This suggests that diagnostic closure is reached by strategies that only reside in their minds because of their skills and experiences. In this research, a number of study techniques were used in order for the information supplied by the system not to conflict with the strategies of the anesthetist.

Szolovits *et al.* [8] states that some approaches to computer-based diagnosis have been based on one of three strategies — flow charts [10], statistical pattern matching [11], or likelihood theory [12]. According to [8], all these three techniques have been successfully applied but each of them has serious hindrances when applied to broad areas of clinical medicine. Flow charts rapidly become unmanageably large and are also unable to deal with cases of uncertainty. Probabilistic and statistical approaches take into account unnecessary assumptions, e.g., that each clinical finding is independent of the other [7]. In Diagnesia, it has been taken into consideration that some disorders are dependant on others thus an attempt to curb some drawback with probabilistic methods.

In Section 1, we discussed the categorization of the input data into five groups: *low*, *low normal*, *normal*, *high normal*, and *high* however, the mapping of the raw values into these categories has not been done. For instance, given a heart rate of 40 bpm, the system cannot tell within which category this value lies. A number of factors need to be considered when making this mapping. A low heart rate for a child may not be low for an old person; a normal rage for a sportsman may not be the same for other persons; and so on. This categorization still needs to be addressed so the system can be integrated into the current monitoring devices to read the values and map them automatically.

Lastly, there are some diagnoses that are completely not in Diagnesia because all their indicators can only be observed and are not measurable. These include haemorrhage (copious discharge of blood from the blood vessels) and hyponatraemia (indicated by arrhythmias and the widening of the QRS complex on the ECG). These diagnoses are never displayed on Diagnesia's output screen and are not counted among the set treated in the system. We therefore still have a challenge of finding a way of indicating such disorders.

Much as we have continued to understand how experts reach diagnostic closure, there has been slow progress towards development of practical decision support programs. Szolovits *et al.* [8] state two major factors that have prevented rapid development in the area. The first one being the huge amount of medical information that has to be gathered even when the medical problem is narrow and the second, "newer cognitive models are so complex that their implementation typically poses a major technical challenge".

Despite this, it is recommended that Diagnesia be tested with a panel of anesthetists and ensure that it displays what is expected of it, and to test whether or not the information provided by the system is of relevancy and enhances the decision making process of the anesthetist. Results obtained here may be compared with those obtained when the first version of the system was tested (see Section 3) in order to have an idea of the progress in the work. The challenges of categorization of the input data and non-measurable indicators (or arguments) have been discussed in this chapter and require attention for further research.

2.7 Acknowledgements

Special thanks go to Constanze Pott of the University of Groningen (RuG), Gerard R. Renaldel de Lavalette (RuG), Albert Ballast, University Medical Center Groningen, Erik Tjong Kim Sang (RuG), and Joost le Feber (RuG) for the tremendous work and efforts that contributed to this chapter.

References

[1] E. Heitmiller, How Anesthesia works. Johns Hopkins Medical School. Retrieved May 2006, from, http://health.howstuffworks.com/anesthesia.htm/printable.

[2] Canadian Anesthesiologists' Society. Retrieved May 2006, from, http://www.cas.ca/anaesthesia/faq/

[3] A. Ballast, Warning Systems in Anesthesia. Rijksuniversiteit Groningen, 1992.

[4] H. H. Bendixen, The tasks of the anesthesiologist. In: Saidman & Smith Eds. Monitoring in Anesthesia. San Diego. Wiley & Sons, 1978.

[5] C. Pott, F. Cnossen, and A. Ballast, More than Psychologists' Chitchat: The Importance of Cognitive Modeling for System Engineering in Anesthesia. University of Groningen, Departments of Computer Science, Artificial Intelligence, and Anesthesiology.

[6] J. le Feber and B. Ballast, Development of a Decision Support System for Anaesthesiologists. University of Groningen. Departments of Mathematics & Computer Science, and Anaesthesiology.

[7] P. Szolovits and S. G. Pauker, Categorical and probabilistic reasoning in medical diagnosis. Artificial Intelligence, 11, 115–144, 1978.

[8] P. Szolovits, Ph.D., S. R. Patil, Ph.D., and B. W. Schwartz, M.D., Artificial Intelligence in Medical Diagnosis. Cambridge and Boston, Massachusetts. Reprinted from Annals of Internal Medicine 108(1), 80–87, January 1988.

[9] A. Tversky and D. Kahneman, Judgment under uncertainty: Heuristics and biases. Science, 185, 1124–1131, 1974.

[10] W. B. Schwartz, Medicine and the computer: the promise and problems of change. N Engl J Med., 283, 1257–1264, 1970.

[11] R. A. Rosati, J. F. Mcneer, C. F. Starmer, B. S. Mittler, J. J. Morris, JR., and A. G. Wallace, A new information system for medical practice. Arch Intern Med., 135, 1017-1024, 1975.

[12] G. A. Gorry, Computer-assisted clinical decision-making. Methods Jnf Med., 12, 45–51, 1973.

3

Development and Testing of a Low Cost, Minimally Invasive Radiofrequency Thermal Probe for Hyperthermia Therapy

Timothy A. Okhai and Cedric J. Smith

Tshwane University of Technology, South Africa

Every year, the burden of benign and malignant tumours continues to increase. So also is the need for better and more effective cancer treatment modalities. The quest for cheaper and more effective treatment methods has led to a growing need to advance the use of thermal ablation therapy as a minimally invasive technique in cancer management. A study was undertaken to investigate the use of radiofrequency (RF) energy in cancer management by developing and testing a minimally invasive thermal probe that will effectively destroy volumes of pathological tumours by means of hyperthermia. This chapter highlights some of the key aspects of this study. The different strategies employed in thermal ablation therapy and the basic principles of RF ablation are briefly discussed. Next, the reaction of tissue to thermal injury at different temperatures is examined. Using experimental tissue samples from soft tissue, kidney, liver, lung and brain tissue, the results of tests carried out with the RF thermal probe to evaluate for effective tissue destruction capability are discussed and evaluated. For each tissue type tested, the findings are evaluated to determine the best power setting for effective tissue ablation over a 15 minutes test period. Finally, the results of these tests are presented and discussed, and a summary of appropriate conclusions arrived at is presented.

3.1 Different Strategies for Thermal Ablation Therapy

Radiofrequency ablation (RFA)

Radiofrequency ablation (RFA) is used to destroy pathological tissue by inducing tissue necrosis through the heating of targeted tissue [1]. While ablation is currently used for the treatment of different diseases, tumour ablation is considered here, i.e. the treatment of cancerous tumours. Apart from RFA, thermal ablation therapy involves other strategies employed in the destruction of cancerous tumours.

Cryoablation therapy (or cryotherapy)

Cryoablation therapy (or cryotherapy) uses liquid nitrogen (or the expansion of argon gas) to freeze and kill abnormal tissue. After numbing the tissue around the mass, a cryoprobe, which is shaped like a large needle, is inserted into the middle of the lesion. An ice ball forms at the tip of the probe and continues to grow until the images confirm that the entire tumour has been engulfed, killing the tissue [2, 3]. The cost of a cryoablation unit ranges upwards from $190,000 and each multi-use probe approximately $3750 [4].

Laser Ablation (or interstitial laser photocoagulation)

Laser Ablation (or interstitial laser photocoagulation) uses a highly concentrated beam of light to penetrate the cancerous tissue. The laser energy is emitted from an optical fibre placed within a needle positioned at the centre of the tumour using either stereotactic guidance or Magnetic Resonance Imaging (MRI) [5], [6]. Two methods for delivery of light have been described to produce larger volumes of necrosis: multiple bare fibres in an array and cooled-tip diffuser fibres. The major drawback to this technique is its cost, requiring $30,000 to $75,000 for a portable, solid-state laser and $3000 per set of multiple (50) user fibres [7].

Microwave ablation (MWA) or microwave coagulation

Microwave ablation (MWA) or microwave coagulation uses microwave tissue coagulator for irradiation. Ultrahigh speed (2450 MHz) microwaves are emitted from a percutaneously placed microwave electrode inserted into the target tissue under ultrasonographic guidance. Microwave irradiation is carried out for about 60 seconds at a power setting of 60W per pulse. During irradiation, the ultrasonographic probe is placed adjacent to the microwave electrode to monitor the effectiveness of the tumour coagulation [8, 9]. A typical microwave generator costs approximately $65,000 [10].

High Intensity Focused Ultrasound (HIFU) ablation

High Intensity Focused Ultrasound (HIFU) ablation is a noninvasive treatment modality that induces complete coagulative necrosis of a deep tumour through the intact skin. HIFU uses sound energy to produce heat [11–13]. Using the principle of radiofrequency ablation therapy, this chapter evaluates the effective tissue destruction capability of a thermal probe applied to soft tissue, kidney, liver, lung and brain tissue at a frequency of 460 kHz and power settings between 1 to 20 watts.

3.2 Principles of Radiofrequency Ablation

Radiofrequency ablation is physically based on radiofrequency current (about 460 kHz) that passes through the target tissue from the tip of an active electrode (RF thermal probe) towards a dispersive electrode which serves as the grounding pad. These two electrodes are connected to a radiofrequency generator. The active electrode has a very small cross-sectional area (a few square millimeters) with respect to the passive electrode. The active electrode is usually fashioned into the form of a needle-like probe that is inserted into the tumour. The dispersive electrode has a much larger area than the active electrode, on the order of 100 cm^2 or larger, and is usually placed firmly behind the right shoulder or the thigh of the subject, depending on the location of the tumour in the body. Current flowing into the dispersive electrode is the same as the current flowing into the active electrode. But since the active electrode has a far smaller cross-sectional area than the dispersive electrode, the current density in amperes per square meter (A/m^2) is far greater. As a result of the difference in current density between the two electrodes, the energy at the tip of the probe leads to ionic agitation with subsequent conversion of friction into heat. Figure 3.1 shows ionic agitation by alternating electric current shown by the sinusoidal waveform.

The tissue ions (represented by the dark and grey circles) are agitated as they attempt to follow the changes in direction of alternating electric current. This change in direction is shown by the arrows pointing in opposite directions. The agitation results in frictional heat around the electrode. The marked discrepancy between the surface area of the needle electrode and the dispersive

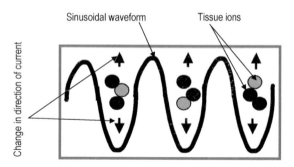

Fig. 3.1 Ionic agitation by alternative electric current [17].

electrode causes the generated heat to be tightly focused and concentrated around the needle electrode. The use of a large grounding pad ensures maximum surface area for dispersion of current from the needle electrode. The grounding pad also maximizes dispersion of equal amounts of energy and heat at the grounding pad sites, thereby minimizes the risk of burns. The tissue underneath the passive electrode heats up only slightly, while the tissue in contact with the active electrode is resistively heated to elevated temperatures sufficient for tumour ablation (coagulative necrosis).

The strategy of RF ablation is to create a closed-loop circuit including the RF generator, the needle electrode, the patient (tissue) and the passive electrode (grounding pad) in series, as shown in Figure 3.2. The heating of tissue is due to the power dissipated in the tissue. Power dissipated in the tissue is determined from the volume of the tissue, the current density at the probe tip, and the resistivity of the tissue.

If I amperes of current flows through V volumes of tissue having a resistivity of ρ, then, the power dissipated in the tissue can be derived from the expression:

$$P = \rho V I_d^2 \tag{3.1}$$

where P is the power in watts (W), ρ is the resistivity of the tissue in Ohm-metres (Ω-m), V is the tissue volume in cubic metres (m^3), and I_d is the current density in amperes per square metre (A/m^2).

Hyperthermia (thermal) coagulation necrosis

Coagulation necrosis denotes "irreversible thermal damage to cells even if the ultimate manifestations of cell death do not fulfill the strict histological criteria

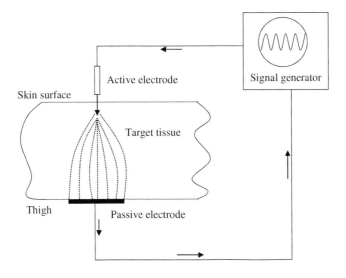

Fig. 3.2 Basic principle of RF ablation system.

of coagulative necrosis" [15]. The nature of the thermal damage caused by radiofrequency heating is dependent on both the tissue temperature achieved and the duration of heating. Events that occur at various temperatures are described as follows [15]:

- At 42° C, cells die but it may take a significant amount of time (approximately 60 min).
- At 42 to 45° C, cells are more susceptible to damage by other agents like chemotherapy and radiation.
- Over 46° C, irreversible damage occurs depending upon duration of heating.
- From 50 to 55° C, duration necessary to shorten irreversible damage to cells is shortened to 4–6 minutes.
- From 50 to 100° C, there is near immediate coagulation of tissue, almost instantaneous protein denaturation, melting of lipid bilayers, irreversible damage to mitochondrial and cytosolic (key cellular) enzymes of the cells, DNA and RNA.
- From 100 to 110° C, tissue vaporizes and carbonizes, all of which decrease energy transmission and impede ablation.

Figure 3.3 shows tissue reaction to thermal injury at different temperatures. For successful ablation to be achieved, the tissue temperature should be maintained in the ideal ablation range to ablate tumour adequately and avoid carbonization around the tip of the electrode due to excessive heating.

For adequate destruction of tumour tissue, the entire volume of a lesion must be subjected to cytotoxic temperatures. Hence effective heating throughout the target volume (i.e. the tumour and about 5 mm thickness around normal tissue) is required, as shown in Figure 3.4. Thus, an essential objective of radiofrequency ablation therapy is achievement and maintenance of 50° C

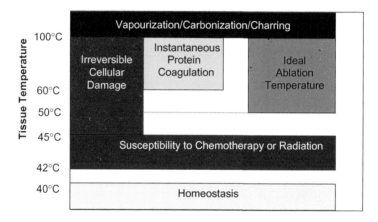

Pathological Reaction of Tissue to Heat

Fig. 3.3 Tissue reaction to thermal injury at different temperatures [17].

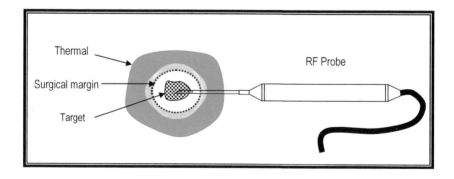

Fig. 3.4 Schematic diagram illustrating RF ablation.

to 100° C temperature throughout the entire target volume for at least 4–6 minutes. However, the relatively slow thermal conduction from the electrode surface through the tissues increases the duration of application to 10–30 minutes.

Recommendations of heating for these extended durations are based on experimental and clinical data suggesting that thermal equilibrium and, hence, complete induction of coagulation are not achieved for a given radiofrequency application until these thresholds are achieved. The development of a radiofrequency ablation system in this study under discussion is aimed at producing a device that is able to satisfy the minimum requirement for effective tumour ablation at the ideal (cytotoxic) ablation temperature.

Appreciable advances have been made over the past decade to produce application devices for radiofrequency ablation. The *Radionics probe* is an internally cooled device that also uses pulsing sequences to improve heating. It is available in one size (17 Ga) and 10, 15, and 25 cm lengths. It comes with a single electrode with a tip exposure of 2–4 cm, or cluster electrode [14]. The *RITA probe* is a 15 Ga device that comes with various arrays. It has a thermocouple at the tip of the probe that registers the tissue temperature, and that is used to monitor its effect [15]. The *LeVeen probe* has multiple (10, 12) tines. There are 2.0, 3.0, 3.5 and 4.0 cm diameter needles from which the tines are deployed. The LeVeen needle electrode is designed to deliver a consistent pattern of heat throughout the lesion [16]. These and other application devices for RFA are available for use in the USA and some parts of Europe. In spite of technical progress in the development of various application devices for radiofrequency ablation therapy, most patients with malignant tumours, especially in Sub-Saharan Africa, have not yet benefitted from this technology due to their limited availability and exorbitant cost. A typical RF generator costs $25,000 and each single use probe costs approximately $800 to $1200 [4]. This chapter presents the structure and experimental results of a low cost minimally invasive radiofrequency thermal probe developed for hyperthermia therapy. The thermal probe developed is effective and economical, and represents more than 70% in cost reduction when compared to commercially available reusable radiofrequency thermal probes reviewed.

3.3 Materials and Methods

The radiofrequency thermal probe developed was designed on a SolidWorks platform and manufactured according to design specifications. The device consists of an RF shielded insulated handle with a needle probe. The shaft of the needle is also insulated except for the tip which makes physical contact with the tumour or volume to be treated. A coaxial cable connects the device to the RF power unit. The RF thermal probe uses a stainless steel needle (size 14G × 3-1/4) with a diameter and length of 2.1 × 80 mm, connected to the conducting coaxial cable in one end, and housed in an epoxy resin holder (probe handle) that is 120 mm long and 15 mm in diameter. The stainless steel needle is insulated, except for the exposed 20 mm tip that makes direct contact with tissue. The insulation prevents normal tissue from being destroyed along with cancerous tissue during thermal ablation treatment. The probe is reusable and is made of epoxy-resin material that can be easily steam-cleaned. An essential objective of radiofrequency ablation therapy is to achieve and maintain a temperature range of 50–100°C throughout the entire target volume for at least 4–6 minutes [17-19]. Figure 3.5 is a schematic drawing of the radiofrequency thermal probe.

From Equation (3.1), power dissipated (P) is directly proportional to volume (V). Tumour is usually treated as a sphere, and volume of a sphere is given by,

$$V = 4/3\pi r^3 \tag{3.2}$$

where r is the radius of the sphere. It follows that, power dissipated is directly proportional to the cube of the radius. The temperature rise follows the

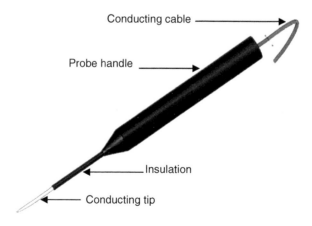

Fig. 3.5 RF thermal probe.

accepted cube root heating function. This means that the outer limit of critical cell temperature where cell necrosis takes place is reasonably well-defined by the applied power and will be spherical around a point source if the impedance remains constant. In practice, we have a short cylindrical contact volume in the tumour with non linear impedances. This results in an egg shaped volume being treated. To verify that the radiofrequency thermal probe developed is a device that is able to satisfy this minimum requirement for effective tumour ablation at the ideal cytotoxic temperature, experimental tests were done with different tissues types to determine how each tissue type responds to RF energy by observing and recording the temperature change at the probe tip. Liver, lung, brain, kidney and soft tissue were tested at different power settings to determine which power setting gives the best results with each tissue type in terms of the minimum time to reach the ideal temperature range, and the maximum time to remain within this range without charring or vapourizing. An RF generator (460 KHz) was connected in a closed circuit with the RF thermal probe, tissue sample, and dispersive electrode in series. Each tissue type was tested with different power settings, and each test was done for 15 minutes.

3.4 Results and Discussion

Both macroscopic and microscopic examination of tissue samples tested show clear evidence of coagulation necrosis as seen with liver tissue in Figure 3.6 before and after ablation. The arrows in Figure 3.6 show the region of tissue destroyed by radiofrequency energy.

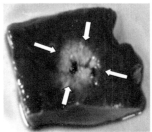

Liver tissue before ablation   Liver tissue after ablation

Fig. 3.6 Macroscopic examination of liver tissue showing evidence of tissue necrosis.

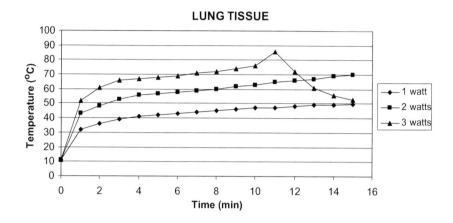

Fig. 3.7 Lung tissue results.

A tissue volume of up to 20 mm diameter was necrosed with this single-tine probe.

The plots of temperature versus time for different tissue types tested using different power settings are presented in the following figures:

From Figure 3.7, it is seen that, while 1 watt was inadequate for coagulation necrosis in lung tissue, 3 watts showed evidence of carbonization, leading to a drop in temperature as further conduction is inhibited. The best result was achieved with 2 watts, which showed a steady rise in temperature maintained within the ideal ablation temperature range.

The plot in Figure 3.8 shows that, while 2 watts was below the ideal temperature range, and therefore inadequate for effective tissue necrosis, 4 watts was too high and showed evidence of carbonization, resulting in a drop in temperature due to inhibition in conduction. The best result in terms of effective tissue necrosis was achieved with the 3 watts power setting.

In Figure 3.9, the plot shows that, while 5 watts and 6 watts are within the ideal ablation zone, the 7 watts setting is too high for liver tissue as it produced carbonization resulting in temperature drop. The best result however was recorded with the 6 watts setting which shows a steady rise in temperature without carbonization or charring.

In Figure 3.10, the results show that 15 watts produced temperature above the ideal ablation range, leading to carbonization and consequently a drop in temperature. Both 10 watts and 13 watts are suitable for ablating kidney tissue

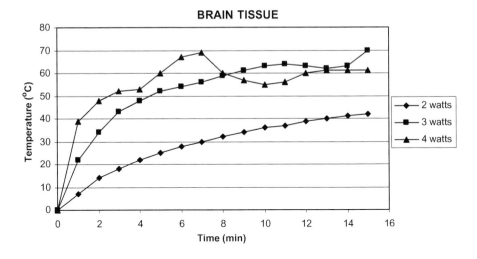

Fig. 3.8 Brain tissue results.

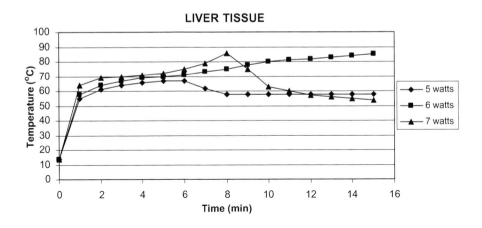

Fig. 3.9 Liver tissue results.

as seen, with 13 watts giving the best results since it allows the use of a higher temperature.

In Figure 3.11, the results show that, while 15 watts produced tempera-ture within the ideal ablation range, the best results was obtained with the 20 watts power setting since the temperature is higher. This means that the ideal temperature range for treatment will be reached quicker with 20 watts.

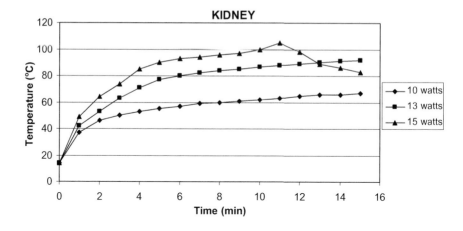

Fig. 3.10 Kidney tissue results.

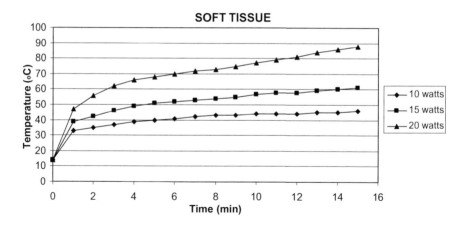

Fig. 3.11 Soft tissue test results.

The 10 watts power setting produces temperature below the ideal range, and was therefore inadequate for ablating soft tissue.

The extent of coagulation necrosis is dependent on three factors: the energy deposited, local tissue interaction minus the heat lost. The relationship between these three factors can be expressed as,

$$Coagulation\ necrosis = energy\ deposited \times local\ tissue\ interactions$$
$$- heat\ loss$$

Heat efficacy is defined as the difference between the amount of heat generated and the amount of heat lost. Therefore, effective tissue ablation can be achieved by optimizing heat generation and minimizing heat loss within the area to be ablated. Heat generation is correlated with the intensity (power output) and duration of the radio-frequency energy deposited. Heat conduction or diffusion is usually explained as a factor of heat loss in regard to the RF thermal probe tip. Heat is lost mainly through convection by means of blood circulation. Therefore, the effect of cooling of tissue by perfusion can limit the reproducible size of the ablation lesion in vivo. However, in the tests performed here, the heat loss is assumed to be negligible because the experimental tissues used for the tests were not in contact with the blood circulation. This minimized the chances of any heat loss by convection, and maximized the amount of heat deposited in the tissue.

3.5 Summary of Results

The above results have been summarized in Table 3.1 showing the different tissue types, the best power settings suitable for each tissue type, the minimum and maximum time required to keep the temperature within the ideal ablation range of 50 to $100°$ C, and the total duration. The total duration is taken as the difference between the maximum time and minimum time.

The economical radiofrequency thermal probe discussed in this chapter was demonstrated to effectively induce tissue necrosis in selected tissue samples. A cost analysis was done to compare the thermal probe with similar probes existing in the overseas market. While reusable probes manufactured overseas cost an average of $3,000.00, the reusable probe presented in this chapter was developed at about 20% of this amount. This is a cost reduction of about 80% (see Table 3.2). In order to compute the real gain in terms of

Table 3.1 Summary of results.

Tissue type	Power (watts)	Min. time (min)	Max. time (min)	Duration (min)
Brain	3.0	4.30	15.0	10.30
Kidney	13.0	1.30	15.0	13.30
Liver	6.0	0.30	15.0	14.30
Lungs	2.0	2.30	15.0	12.30
Soft tissue	20.0	1.30	15.0	13.30

Table 3.2 Cost analysis for local manufacture of RF probe.

Quantity	Unit	Unit Price (x) (14% VAT incl.)	Approximate Cost* (y) $[(x/3000) \times 100]$ %	Approximate Cost Reduction** $(100 - y)$ %
1	1	$585.20	20.00	80.00
10	1	$456.00	15.00	85.00
50	1	$410.70	14.00	86.00
100	1	$369.94	12.00	88.00
500	1	$333.24	11.00	89.00
1000	1	$300.23	10.00	90.00

*Percentage of production cost when compared with overseas equivalent.
**Percentage cost reduction when compared with overseas equivalent.

cost-effectiveness of the probe developed, a quotation for the manufacture of different quantities (10, 50, 100, 500 and 1,000) of the thermal probe was obtained from a manufacturing firm. Based on the quotation provided, a cost analysis was done to compare the unit cost of producing different quantities (including the single probe), with the cost of a similar (re-useable) probe available in the overseas market at about $3,000.00 The result of this cost analysis is presented in Table 3.2.

From Table 3.2 above, for a single probe produced, a percentage cost reduction of up to 80% is achievable. If 100 probes are produces, each unit will cost 12% of the overseas cost to produce. This is a cost reduction of about 88%. For 1000 units, it will cost only 10% of the overseas cost to produce. This means that a percentage cost reduction of up to 90% and above is achievable with mass production of the RF thermal probe. Even though the probe developed is just a prototype, this research work has demonstrated that the development of a minimally invasive oncology probe that will drastically reduce cost and level of trauma to patients is a real possibility that should be explored further.

3.6 Summary of Chapter

In this chapter the various strategies for thermal ablation therapy were presented. They include cryoablation, laser ablation, microwave ablation, high intensity focused ultrasound ablation, and radiofrequency ablation. The basic principles of radiofrequency ablation for hyperthermia therapy were discussed in detail. To induce tissue necrosis in hyperthermia therapy, the key aim is to achieve and maintain a temperature range of 50 to 100°C throughout the entire tissue volume for 4 to 6 minutes [19]. To determine if this objective was achieved, the RF thermal probe developed was tested with different tissue samples. Both the test results and histological analysis of tissue samples tested conclusively show that the RF thermal probe developed is an effective, economical, minimally invasive probe for hyperthermia therapy.

Problems

Problem 3.1

Thermal ablation therapy employs different strategies in managing cancerous tumours. Mention these strategies and briefly describe each one.

Problem 3.2

Use a labeled diagram to explain the basic principle of radiofrequency ablation. Why is it important to use a dispersive electrode with a large surface area as the grounding pad?

Problem 3.3

What is the ideal ablation temperature range for radiofrequency ablation? Briefly discuss the effects of carbonization and charring on tissue conduction during hyperthermia therapy.

Problem 3.4

The heating of tissue during hyperthermia therapy is due to the power dissipated in the tissue. Calculate the power dissipated in 0.2 m^3 of tissue that has a resistivity of $1.6 \times 10^3 \Omega$-m if the current density is 0.18 A/m^2.

Problem 3.5

Calculate the expected temperature 15 mm from the centre of the tumour if the temperature measured 8 mm from the probe is 55°C.

Problem 3.6

Draw a graph of the expected times for full cell necrosis versus temperature.

Problem 3.7

Compare the physiological effects of hyper thermal and hypo thermal cell destruction.

References

[1] S.E. Singleton, "Radiofrequency ablation of breast cancer", *American journal of surgery*, vol. 69, pp. 37–40, 2003.

[2] B.D. Cowen, P.E. Sewell, J.C. Howard, R.M. Arriola, and L.G. Robinette, "Interventional magnetic resonance imaging cryotherapy of uterine fibroid tumours: Preliminary observation", *American journal of obstetricians and gynecologists*, vol. 186, no. 6, pp. 1183–1187, 2002.

[3] Anonymous, "Ablation therapy destroys breast cancer without scarring", *Radiological Society of North America*, [Online], Available from: http://www.rsna.org/, 2002, [Last accessed: 10 June 2008].

[4] G.D. Dodd III, M.C. Soulen, and R.A. Kante *et al.*, "Minimally invasive treatment of malignant hepatic tumors: At the threshold of a major breakthrough", *Radiographics, 2000*, vol. 20, pp. 9–27.

[5] K.J. Bloom, K. Dowlat, and L. Assad, "Pathologic changes after interstitial laser therapy of infiltrating breast carcinoma", *American journal of surgery*, vol. 182, pp. 384-388, 2001.

[6] M.S. Sabel, "In Situ Ablation of Breast Tumors. What is The State of the Art?", *Cancernews*, [Online], Available from: http://www.cancernews.com/articles/breastcancer therapies.htm, 2001, [Last accessed: 19 August 2008].

[7] A. Shah, "Recent developments in the chemotherapeutic management of colorectal cancer", *BC Medical Journal*, vol. 42, pp. 180–182, 2000.

[8] R.A. Gardner, H.I. Vargas, and J.B. Block, "Focused microwave phased array thermotherapy for primary breast cancer", *Annals of surgical oncology*, vol. 9, pp. 326–332, 2002.

[9] T. Ishikawa, T. Kohno, T. Shibayama, Y. Fukushima, T. Obi, T. Teratani,, S. Shiina, Y. Shiratori, and M. Omata, "Thoracoscopic thermal ablation for hepatocellular carcinoma located beneath the diaphragm", *Endoscopy*, vol. 33(8), pp. 697–702, 2001.

[10] S.G. Ho, P.L. Munk, G.M. Legiehn, S.W. Chung, C.H. Scudamore, and M.J. Lee, "Minimally invasive treatment of colorectal cancer metastasis: Current status and new directions", *BC Medical Journal*, vol. 42, no. 10, pp. 461–464, 2002.

[11] K. Hynynen, O. Pomeroy, and D.N. Smith, "MR Imaging-guided focused ultrasound surgery of fibroadenomas in the breast: a feasibility study", *Radiology*, vol. 219, pp. 176–185, 2001.

[12] F. Wu, Z.B. Wang, Y.D. Cao, W.Z. Chen, J. Bai, J.Z. Zou, and H. Zhu, "A randomized clinical trial of high-intensity focused ultrasound ablation for the treatment of patients with localized breast cancer", *British journal of cancer*, vol. 89, pp. 2227–2233.

[13] F. Wu, Z. Wang, W. Chen, H. Zhu, J. Bai, J. Zou, K. Li, C. Jin, F. Xie, and H. Su, "Extracorporeal high intensity focused ultrasound ablation in the treatment of patients with large hepatocellular carcinoma", *Annals of surgical oncology*, vol. 11, pp. 1061–1069, 2004.

[14] Anonymous, *RadiologyInfo*, [Online], Available from: http://www.radiologyinfo.org/en/photocat/photos_pc.cfm?image=ri-rfa-devices.jpg&pg=rfa, 2007, [Last accessed: 15 September 2008].

[15] Caridi, J. (comp.), "Radio-frequency ablation", University of Florida, Florida, 2004.

[16] Anonymous, *Boston Scientific Company,* [Online], Available from: http://www.bostonscientific.com/med_specialty/deviceDetail.jsp?task=tskBasicDevice.

jsp&sectioned=4&reIId=4,178,179,180 &deviceId=13004, 2007, [Last accessed: 27 June 2008].

[17] H. Rhim, S.N. Goldberg, G.D. Dodd, L. Solbiati, K.L. Lim, M. Tonolini, and O.N. Cho, "Helping the hepatic surgeon: Essential techniques for successful radio-frequency thermal ablation of malignant hepatic tumours", *Radiographics*, vol. 21, pp. S17–S35, 2007.

[18] S.N. Goldberg, L. Solbiati, G.S. Gazelle, K.K. Tanabe, C.C. Compton, and P.R. Mueller, "Treatment of intrahepatic malignancy with radio-frequency ablation: radiologic-pathologic correlation in 16 patients" (abstr), *American journal of roentgenology, 168. [American Roentgen Ray Society 97th Annual Meeting Program Book suppl]*, 1997, pp. 121.

[19] S.N. Goldberg, G.S. Gazelle, and P.R. Mueller, "Thermal ablation therapy for focal malignancy. A unified approach to underlying principles, techniques, and diagnostic image guidance", *American journal of roentgenology*, 2000, vol. 174, pp. 323-331.

4

Comparative Functional Magnetic Resonance Imaging Using Brain Imaging Modalities

Mohammad Karimi Moridani

Islamic Azad University, Tehran, Iran

4.1 Introduction

Functional brain imaging using MRI (functional MRI or fMRI) has become a valuable tool for studying function/structure relationships in the human brain in both normal and clinical populations. This paper describes the physiological changes associated brain with activity, including changes in blood flow, volume, and oxygenation. The latter of these, known as Blood Oxygenation Level Depended (BOLD) contrast, is the most common approach for functional MRI, but it is related to brain activity via a variety of complex mechanisms. Blood oxygenation level dependent functional MRI and near infrared optical tomography have been widely used to investigate hemodynamic responses to functional stimulation in the human brain. The temporal hemodynamic response shows an increased total hemoglobin concentration, which indicates an increased cerebral blood volume (CBV) during physiological activation. Blood Oxygenation Level Dependent signal indirect measure of neural activity. The signal variations induced by respiration and cardiac motion decrease the statistical significance in functional MRI data analysis. Physiological noise can be estimated and removed adaptively using signal projecting technique with the actual functional signal preserved. We estimate and remove the physiological noise from the magnitude images. This method is a fully data-driven method, which can efficiently and effectively reduce the overall signal fluctuation of functional MRI data. Assumes that the MRI data recorded on each trial are composed of a signal added with noise Signal (random) is present on every trial, so it remains constant through averaging and Noise randomly varies across trials, so it decreases with averaging Thus, Signal-to-Noise Ratio (SNR) increases with averaging.

4.2 Comparison of the Functional Brain Imaging Modalities

The brain imaging techniques that had been presented in this paper all measure slightly different properties of the brain as it carries out cognitive tasks. Because of this the techniques should be seen as complementary rather than competitive. All of them have the potential to reveal much about the function of the brain and they will no doubt develop in clinical usefulness as more about the underlying mechanisms of each are understood, and the hardware becomes more available. A summary of the strengths and weaknesses of the techniques is presented in Table 4.1 [5].

4.2.1 SPECT and PET

The imaging modalities of single photon emission computed tomography (SPECT) and positron emission tomography (PET) both involve the use of radioactive nuclides either from natural or synthetic sources. Their strength

Table 4.1 Comparison of modalities for studying brain function.

Technique	Resolution	Advantages	Disadvantages
SPECT	10 mm	Low cost Available	Invasive Limited resolution
PET	5 mm	Sensitive Good resolution Metabolic studies Receptor mapping	Invasive Very expensive
EEG	poor	Very low cost Sleep and operation monitoring	Not an imaging technique
MEG	5 mm	High temporal resolution	Very Expensive Limited resolution for deep structures
fMRI	3 mm	Excellent resolution Non-invasive	Expensive Limited to activation studies
MRS	low	Non-invasive metabolic studies	Expensive Low resolution

is in the fact that, since the radioactivity is introduced, they can be used in tracer studies where a radiopharmaceutical is selectively absorbed in a region of the brain. The main aim of SPECT as used in brain imaging, is to measure the regional cerebral blood flow (rCBF). The earliest experiments to measure cerebral blood flow were performed in 1948 by Kety and Schmidt [1]. They used nitrous oxide as an indicator in the blood, measuring the differences between the arterial input and venous outflow, from which the cellular uptake could be determined.

This could only be used to measure the global cerebral blood flow, and so in 1963 Glass and Harper [2], building on the work of Ingvar and Lassen, used the radioisotope Xe-133, which emits gamma rays, to measure the regional cerebral blood flow. The development of computed tomography in the 1970's allowed mapping of the distribution of the radioisotopes in the brain, and led to the technique now called SPECT [3].

4.2.2 EEG and MEG

Measuring the electrical signals from the brain has been carried out for several decades [4], but it is only more recently that attempts have been made to map electrical and magnetic activity. The electroencephalogram (EEG) is recorded using electrodes, usually silver coated with silver chloride, attached to the scalp and kept in good electrical contact using conductive electrode jelly. One or more active sites may be monitored relative to a reference electrode placed on an area of low response activity such as the earlobe. The signals are of the order of 50 microvolts, and so care must be taken to reduce interference from external sources, eye movement and muscle activity. Several characteristic frequencies are detected in the human EEG. For example, when the subject is relaxed the EEG consists mainly of frequencies in the range 8 to 13 Hz, called alpha waves, but when the subject is more alert the frequencies detected in the signal rise above 13 Hz, called beta waves. Measurements of the EEG during sleep have revealed periods of high frequency waves, known as rapid eye movement (REM) sleep which has been associated with dreaming [5].

MEG experiments are carried out in much the same way as their EEG counterpart. Having identified the peak of interest, the signals from all the detectors are analysed to obtain a field map. From this map an attempt can be made to ascertain the source of the signal by solving the inverse problem.

Since the inverse problem has no unique solution, assumptions need to be made, but providing there are only a few activated sites, close to the scalp then relatively accurate localization is possible, giving a resolution of the order of a few millimeters.

MEG has the advantage over EEG that signal localisation is, to an extent, possible, and over PET and fMRI in that it has excellent temporal resolution of neuronal events. However MEG is costly and its ability to accurately detect events in deeper brain structures is limited.

4.2.3 Functional MRI and MRS

Since functional magnetic resonance imaging (fMRI) is the subject of this paper, little will be said in this section as to the mechanisms and applications of the technique. The purpose of this section is to compare fMRI to the other modalities already mentioned, and also to consider the related, but distinct technique of magnetic resonance spectroscopy (MRS) [5].

During an fMRI experiment, the brain of the subject is scanned repeatedly, usually using the fast imaging technique of echo planar imaging (EPI). The subject is required to carry out some task consisting of periods of activity and periods of rest. During the activity, the MR signal from the region of the brain involved in the task normally increases due to the flow of oxygenated blood into that region. Signal processing is then used to reveal these regions. The main advantage of MRI over its closest counterpart, PET, is that it requires no contrast agent to be administered, and so is considerably safer. In addition, high quality anatomical images can be obtained in the same session as the functional studies, giving greater confidence as to the source of the activation. However, the function that is mapped is based on blood flow, and it is not yet possible to directly map neuroreceptors as PET can. The technique is relatively expensive, although comparable with PET, however since many hospitals now have an MRI scanner the availability of the technique is more widespread [5].

FMRI is limited to activation studies, which it performs with good spatial resolution. If the resolution is reduced somewhat then it is also possible to carry out spectroscopy, which is chemically specific, and can follow many metabolic processes. Since fMRS can give the rate of glucose utilisation, it provides useful additional information to the blood flow and oxygenation measurements from fMRI, in the study of brain metabolism [5].

4.3 Metabolism and Blood Flow in the Brain

The biochemical reactions that transmit neural information via action potentials and neurotransmitters, all require energy.

This energy is provided in the form of ATP, which in turn is produced from glucose by oxidative phosphorylation and the Kreb's cycle (Figure 4.1) [5].

As ATP is hydrolysed to ADP, energy is given up, which can be used to drive biochemical reactions that require free energy. The production of ATP from ADP by oxidative phosphorylation is governed by demand, so that the energy reserves are kept constant. That is to say, the rate of this reaction depends mainly on the level of ADP present.

This means that the rate of oxygen consumption by oxidative phosphorylation is a good measure of the rate of use of energy in that area [5].

The oxygen required by metabolism is supplied in the blood. Since oxygen is not very soluble in water, the blood contains a protein that oxygen can bind to, called haemoglobin. The important part of the haemoglobin molecule is an iron atom, bound in an organic structure, and it is this iron atom which gives blood it's colour. When an oxygen molecule binds to haemoglobin, it is said to be oxyhaemoglobin, and when no oxygen is bound it is called deoxyhaemoglobin.

To keep up with the high energy demand of the brain, oxygen delivery and blood flow to this organ is quite large. Although the brain's weight is only 2% of the body's, its oxygen consumption rate is 20% of the body's and blood flow 15%. The blood flow to the grey matter, which is a synapse rich area, is about 10 times that to the white matter per unit volume. Regulation of the

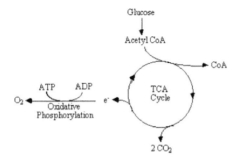

Fig. 4.1 Overview of the aerobic metabolism of glucose to ATP following the Kreb's cycle.

regional blood flow is poorly understood, but it is known that localised neural activity results in a rapid selective increase in blood flow to that area [5].

4.4   Blood Oxygen Level Dependent Contrast in MR Images

Since regional blood flow is closely related to neural activity, measurement of the rCBF is useful in studying brain function. It is possible to measure blood perfusion with MRI, using techniques similar to those mentioned. However there is another, more sensitive, contrast mechanism which depends on the blood oxygenation level, known as blood oxygen level dependent(BOLD) contrast. The mechanisms behind the BOLD contrast are still to be determined, however there are hypotheses to explain the observed signal changes.

Deoxyhaemoglobin is a paramagnetic molecule whereas oxyhaemoglobin is diamagnetic. The presence of deoxyhaemoglobin in a blood vessel causes a susceptibility difference between the vessel and its surrounding tissue. Such susceptibility differences cause dephasing of the MR proton signal [6], leading to a reduction in the value of $T2^*$. In a $T2^*$ weighted imaging experiment, the presence of deoxyhaemoglobin in the blood vessels [7, 8] causes a darkening of the image in those voxels containing vessels. Since oxyhaemoglobin is diamagnetic and does not produce the same dephasing, changes in oxygenation of the blood can be observed as the signal changes in $T2^*$ weighted images [5].

It would be expected that upon neural activity, since oxygen consumption is increased, that the level of deoxyhaemoglobin in the blood would also increase, and the MR signal would decrease. However what is observed is an increase in signal, implying a decrease in deoxyhaemoglobin. This is because upon neural activity, as well as the slight increase in oxygen extraction from the blood, there is a much larger increase in cerebral blood flow, bringing with it more oxyhaemoglobin (Figure 4.2). Thus the bulk effect upon neural activity is a regional decrease in paramagnetic deoxyhaemoglobin, and an increase in signal [5].

The study of these mechanisms are helped by results from PET and near-infrared spectroscopy (NIRS) studies. PET has shown that changes in cerebral blood flow and cerebral blood volume upon activation, are not accompanied by any significant increase in tissue oxygen consumption [9].

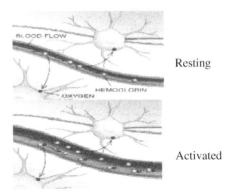

Resting

Activated

Fig. 4.2 Upon activation, oxygen is extracted by the cells, thereby increasing the level of deoxyhaemoglobin in the blood. This is compensated for by an increase in blood flow in the vicinity of the active cells, leading to a net increase in oxyhaemoglobin.

NIRS can measure the changes in concentrations of oxy- and deoxy-haemoglobin, by looking at the absorbency at different frequencies. Such studies have shown an increase in oxyhaemoglobin, and a decrease deoxy-haemoglobin upon activation. An increase in the total amount of haemoglobin is also observed, reflecting the increase in blood volume upon activation [2].

4.5 Conclusion

Functional magnetic resonance imaging can accurately represent cerebral topography, cortical venous structures and underlying lesions. Functional activation appears to accurately localize appropriate cortical areas and these studies are feasible in the presence of local pathology. It is extremely useful in presurgical planning as well as assessment of operability. Intra-operatively, it shows a great promise in being able to define the exact location and extent of lesions with respect to surrounding functional cortex.

Functional MRI is a new and powerful neuroimaging technique that can create an anatomical and functional model of an individual patient's brain. The concurrent 3-D rendering of cerebral topography, cortical veins and related pathology gives an unprecedented display of critical relational anatomy. Since stereotaxy means simply the three dimensional arrangement of objects, then fMRI may be the ultimate stereotactic system. It allows us to see through the scalp and cortex into subcortical areas which are not visually apparent.

It accurately predicts cortical gyral and venous anatomy as well as the subcortical location and extent of lesions. But most importantly, it is capable of mapping specific cortical functions to anatomical regions thereby combining form and function.

We have described the physiological bases of functional MRI and introduced a physiologically relevant model of the vascular response to fMRI. We described the major optimization goals in fMRI and several fMRI acquisition approaches. Substantial progress is being made to reduce artifacts in fMRI as well as to improve the measurement of alternate physiological phenomena using MRI.

The spatial activation pattern of changes indeoxyhemoglobin concentration is consistent with the BOLD signal map. The patterns of oxy- and deoxyhemoglobin concentrations are very similar to one another. The temporal hemodynamic response shows an increased total hemoglobin concentration, which indicates an increment of cerebral blood volume (CBV) during physiological activation. It has now been firmly established that magnetic resonance imaging can be used to map brain function.

The main impetus of research and development of the technique, needs to be directed in several areas if fMRI is to become more than 'colour phrenology', intriguing in its results yet having little clinical value. The mechanisms behind the BOLD effect need to be better understood, as does the physiological basis of the observed blood flow and oxygenation changes. The combination of the functional imaging modalities needs attention, since it is unlikely that any one method will provide the full picture. Finally, robust and simple techniques for data analysis need to be developed, allowing those who do not specialise in fMRI, to carry out experiments and interpret results.

References

[1] S. S. Kety, and C. F. Schmidt, (1948). "The Nitrous Oxide Method for the Quantitative Determination of Cerebral Blood Flow in Man: Theory, Procedure and Normal Values." *J. Clin. Invest.* 27, 476–483.

[2] H. I. Glass, and A. M. Harper, (1963). "Measurement of regional blood flow in cerebral cortex of man through intact skull." *Br. Med. J.* 1, 593.

[3] D. E. Kuhl, and R. Q. Edwards, (1963). "Image separation radioisotope scanning." *Radiology* 80, 653–661.

[4] H. Berger, (1929). "Über das elektrenkephalogramm des menschen." *Arch. Psychiatr Nervenkr* 87, 527–570.

[5] veopen. (1995), "Magnetic resonance imaging of brain function." *Magn. Reson. Med.* 22, 149–166.

[6] K. R. Thulborn, J. C. Waterton, P. M. Matthews, and G. K. Radda, (1982). "Oxygen dependence of the transverse relaxation time of water protons in whole blood at high field." *Biochim. Biophys. Acta.* 714, 265–270.

[7] S. Ogawa, T. M. Lee, A. S. Nayak, and P. Glynn, (1990). "Oxygenation-sensitive contrast in magnetic resonance image of rodent brain at high magnetic fields." *Magn. Reson. Med.* 14, 68–78.

[8] S. Ogawa and T. M. Lee, (1990). "Magnetic resonance imaging of blood vessels at high fields: *In Vivo* and *in Vitro* measurements and image simulation." *Magn. Reson. Med.* 16, 9–18.

[9] P. T. Fox, M. E. Raichle, M. A. Mintun, and C. Dence, (1988). "Nonoxidative glucose consumption during physiologic neural activity." *Science* 241, 462–464.

5

Design of a Neural Network Classifier for Separation of Images with One Chromosome from Images with Several Chromosomes

Yaser Rahimi*, Rassoul Amirfattahi† and Reza Ghaderi*

*Mazandaran University, Iran
†Isfahan University of Technology, Isfahan

5.1 Introduction

The body of each life is organized from cells. Each cell has a nucleus. In the nucleus there is a very large molecule that is named DNA (Deoxy Ribonucleic Acid). This molecule is critically important and determines all properties and behavior of life. In normal form this very large and complex molecule is in the nucleus liquid and can not be viewed with a common microscope. At cell division this molecule is compressed to a shape that is called chromosome.

One of the properties of chromosome is highly colorable. If we colorize the cell, at the time of cell division, in a stage of cell division called metaphase, chromosomes are viewed as stains like X. Naturally human has 23 pairs of chromosomes. Each chromosome has 4 arms. The crossing point of arms is called centromere.

The relationship between properties of life and chromosomes, and the fact that the origin of some diseases is abnormalities in chromosomes, are two reasons to pay attentions to analysis of chromosome shape for detection of properties of life and origins of some diseases [2]. Shape analysis of chromosomes is performed manually by human expert.

Manual karyotyping is slow, and needs a cytogenetics expert. In some diseases like leukemia fast response of test is needed. Also karyotyping is expensive. For these reasons we seek on a system that automatically performs karyotyping using image processing and pattern recognition techniques.

The goal of automatic karyotyping is creating a system that automatically acquires images of chromosomes and after processing and analysis, present a karyotype image and also help genetic experts in detecting abnormalities.

In automatic karyotyping, when we separate chromosomes from image as an object, some chromosomes that are touched or overlapped is detected as an object. For preparing a karyotype and other processing, we must have one chromosome in each image. Images having more than one chromosome, need further processing. So after separation of chromosomes, images with two or more chromosomes must be segregate from that having one chromosome, and processed separately. In this paper we want to design a classifier that automatically separates one-chromosome images from multichromosome images. In Figure 5.1 you see two examples of one-chromosome images, and two examples of multi-chromosome images.

In the context of chromosome image processing [17, 18], chromosome classification [10], detection of specifications of chromosomes [13], processing on touching chromosomes [4,14], processing on overlapping chromosomes [4, 5, 6, 14], processing on FISH[1] images [11, 12, 16], many papers have published, but about automatic separation of multichromosome images from one-chromosome images no research has been reported. Several papers discussing about processing of images with two or more chromosomes, suppose that these images has been detected before. So no solution has been reported for this problem.

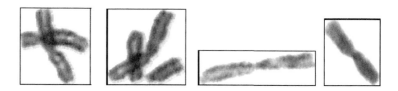

Fig. 5.1 One-chromosome images (right) and multi-chromosome images (left).

---

[1]Florescent In-Situ Hybridization

5.2 Karyotype

For viewing chromosomes, samples of human cells (usually blood cells) are obtained and planted in laboratory and prepare for cell division. Starting cell division, the cell is stained with materials like Giemsa [1]. Then in the stage that chromosomes have maximum clarity (metaphase stage), cell division process is stopped and an image is acquired.

After imaging the chromosomes, the first step of study on chromosomes, is classification of them according to standard pattern named karyotype [8]. Classification is performed on the basis of their length, shape and location of centromere [8].

Karyotype is an abstracted formula for showing the genetic structure of a life. Then this image is studied and each chromosome is compared with standard pattern, recognizing properties and genetic abnormalities of chromosome [3]. The standard image is called ideogram.

5.3 Preprocessing

Images processed in this research, are obtained in Isfahan Center of Medical Genetics, In Isfahan, Iran, by Dr. Valian Borujeni. One of these images is shown in Figure 5.2. In the first step of preprocessing, for reduction of computations, captured color image is converted to gray-scale image. This conversion has no effect on existing information in image, and don't reduce information of image.

As you see in Figure 5.2, the background has many noisy spots and other particles are viewed in image. For separating chromosomes, background must be monotonic. Therefore histogram of image is scaled until noisy points are eliminated and background has maximum difference in histogram with chromosomes. The resulting image is shown in Figure 5.3.

In improved image, although chromosomes are very clear than before, there are some additional spots in image. Also analyzing all images, we recognize that some of spots exist in all images with the same size and location. These spots are related to image acquisition system. Detecting these spots, we make an image containing only spots, and add the negative of this image to improved image. Therefore the troublous spots are deleted. Now, the image is ready for separating chromosomes.

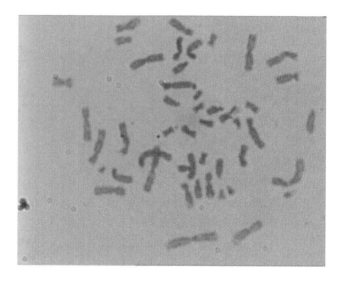

Fig. 5.2 The original image, acquired in Isfahan Center of Medical Genetics, Isfahan, Iran.

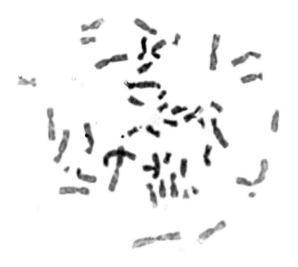

Fig. 5.3 The improved image.

Before separating chromosomes, first convert it to a binary image. The threshold of conversion is obtained from histogram of image.

For separating chromosomes, we first detect all objects in image and number them. In this step even small particles and other objects in image is detected. Then find the location of image and calculate number of pixels in each object. Very large and very small particles are not chromosomes, so these objects are neglected. The result of this step is several images, that there is one object in each image. Number of these images depends on the number of objects in original image. Obtained images are enhanced to easy next processing. Enhancements are as follows:

1. Deletion of very small and very large objects as non-chromosome images.
2. Deletion of parts of other chromosomes that is remained in image.
3. Perfectly whitening the background.

Now separated images are ready. Some of them have more than one chromosome that must be segregate from one-chromosome images. The goal of this research is that these two types of images are automatically separated by a classifier.

5.4 Existed Ideas

First stage of separation is identifying a feature shape for a single chromosome. For computer, touching chromosomes and overlapping chromosomes is not defined. Also in attention to the fact that chromosomes are very soft objects, can be in any form, we can't define any pattern. Only the specification that can be defined for a chromosome is that it is an object with four arms that often isometric arms, are touched together and is viewed as two arms connecting at a point (centromere).

Analyzing multi-chromosome and one-chromosome images, we find these differences between them: first, the dimensions of multi-chromosome images is often larger than one-chromosome images, and second the ratio of surface of chromosomes to surface of background is larger for multi-chromosome images.

5.5 Neural Network Classifier

The goal of this research is designing a classifier that its input is an image and its output is decision of being multi-chromosome or one-chromosome image. Since the gray-scale level of image has no effect on decision of human, we can convert images to binary. The output of classifier has two conditions that we can show with one bit. Therefore, if the classifier output is "zero," the image is one-chromosome and if it is "one," the image is multi-chromosome image.

If we want to use a classifier except neural network classifier, we should define a procedure for feature extraction. But if we use neural network classifier, we don't need to define features for neural network. When a neural network is learning, in fact it learns the features. In this condition, decision space is divided in two regions. Decision boundary is specified during learning process.

5.6 Tests

The data set used for test, has 178 images, which these images are obtained from processing on primary images and separation of chromosomes. These images are different in dimensions. 69 images are multichromosome images and other (109 images) are one chromosome images. 100 images are used for learning of neural network and remaining 78 images are used for test, which 51 images are multi-chromosome and other are one-chromosome images.

The feedforward multi-layer perceptron is the selected neural network. The reason for selecting this neural network is that in previous researches in the context of chromosome image processing and chromosome classification, MLP neural network is used and researches prefer this type of neural network than other neural networks and other classifiers [9].

5.6.1 First Simulation

The shape of each chromosome can be explained with statistical momentums such as average, variance and higher momentums [7]. The benefit of using these momentums is that momentums are extracted features that are independent of rotation (R), scaling (S) and transport (T) [15]. In the first simulation, inputs of neural network were RST-invariant momentums introduced in [15]. After calculating all of momentums, it is discovered that the seventh momentum is zero for all images. So we didn't include it. Therefore the inputs of network are six momentums, and the input layer has six neurons. The second layer is considered as variable. Since one neuron is sufficient for representing two classes, the output layer has one neuron. The output of neural network is thresholded.

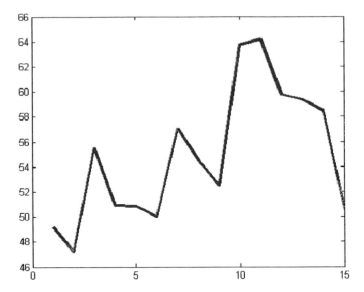

Fig. 5.4 The results of neural network simulation with 6 features at input. Horizontal axis is the number of neurons in hidden layer and vertical axis represnts the average percentage of true classification.

The sufficient memory is required in the network, to represent fundamental generalization with these features, so it is better that the second layer has several neurons.

The simulation was performed with different number of neurons in second layer. For each case, the neural network is learned 20 times and then tested. The results are averaged and, the average is reported as final result of the network. The average results can be viewed in Figure 5.4.

According to Figure 5.4 the best result is 64% and obtained with 11 neurons in hidden layer. So it is recommended that the number of neurons in hidden layer be 11 neurons. Existence of few neurons in hidden layer results that the network can't save required information.

5.6.2 Second Simulation

In the second simulation, features that is considered as inputs of neural network classifier include: surface of image, surface of chromosome , number of pixels of boundary of chromosomes, and six momentums applied in the first simulation. Number of neurons in input layer is equal to the number of inputs, so the input layer has 9 neurons. Output layer of the neural network has one

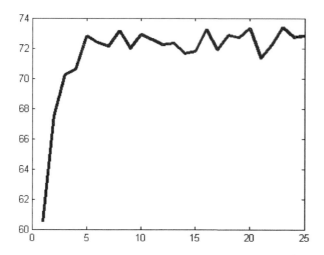

Fig. 5.5 The results of neural network simulation with 9 features at input. Horizontal axis is the number of neurons in hidden layer and vertical axis represents the average percentage of true classification.

neuron, similar to the first simulation. Number of neurons in hidden layer is considered as variable. Simulation of neural network is performed with different neurons in hidden layer, and each simulation is repeated 30 times. The average of true separation percentage is shown in Figure 5.5.

Figure 5.5 shows that the best result is 73% and obtained with 8, 16, 20 and 23 neurons in hidden layer. Figure 5.5 shows another property: When the number of neurons in hidden layer is increased from a threshold (5 neurons), increasing it has no meaningful effect on the result.

5.7 Conclusion

We select a neural network that has the best result. Best result is obtained in second simulation with 73% true separation. The specifications of this neural network are: multi-layer feedforward perceptron (MLP) neural network with one hidden layer that has 9 neurons in input layer, 8 neurons in hidden layer and one neuron in input layer. Input vector of neural network includes: surface of image, surface of chromosome, number of pixels of boundary of chromosomes, and six momentums.

Since no article has published in this context, we can not compare obtained results with other. The results of this research can be used in designation of full-automatic karyotype system.

5.8 Acknowledgement

We thank to Dr. Valian Borujeni, professor of genetic group in Isfahan University, and presidency of Isfahan Center of Medical Genetics for preparing images of chromosomes.

References

[1] A. Arzani, *Genetics and Cytogenetics Laboratort Help*, Arkan Isfahan, Iran, 1996.

[2] M.V. Tompson and J.S. Tompson, *Genetics in Medicine*, (translated to persian) Yavardaran Publication, Iran, 1991.

[3] A. Haji Beigi and R. Radpur, *Educational Dictionary of Genetic Terms*, Hayan- Aba Saleh Publication, Iran, 2002.

[4] G. Agam and I. Dinstein, "Geometric separation of partially overlapping nonrigid objects applied to automatic chromosome classification," *IEEE Transactions On Pattern Analysis And Machine Intelligence*, Vol. 19, No. 11, November 1997, pp. 1212–1222, 1997.

[5] G.C. Charters and J. Graham, "Disentangling chromosome overlaps by combining trainable shape models with classification evidence," *IEEE Transactions On Signal Processing*, Vol. 50, No. 8, August 2002 — pp. 2080–2085, 2002.

[6] G.C. Charters and J. Graham, *Trainable grey-level models for disentangling overlapping chromosomes*, Elsevier-Pattern Recognition 32, pp. 1335–1349, 1999.

[7] R.C. Gonzalez and R.E. Woods, *Digital Image Processing*, 2nd Edition, Prentice-Hall, USA, 2002.

[8] A.J.F. Griffiths, S. R. Wessler, R.C. Lewontin, W.M. Gelbart, D.T. Suzuki, J. H. Miller, *An Introduction to Genetic Analysis*, 2004.

[9] B. Lerner, H. Guterman, I. Dinstein, Y. Romem, "A comparison of multilayer perceptron neural network and bayes piecewise classifier for chromosome classification," *Proceedings of IEEE Conference on Neural Network- part 6 —* pp. 3472–3477, 1994.

[10] B. Lerner and N.D. Lawrence, "A comparison of state-of-the-art classification techniques with application to cytogenetics," *Neural Computing & Applications*, vol. 10(1), pp. 39–47, 2001.

[11] B. Lerner, *Bayesian fluorescence in-situ hybridization signal classification*, Artificial Intelligence in Medicine, vol. 30(3), A special issue on Bayesian Models in Medicine, pp. 301–316, 2004.

[12] B. Lerner, W. F. Clocksin, S. Dhanjal, M. A. Hult'en, and C. M. Bishop, "Automatic signal classification in fluorescence *in-situ* hybridization images," *Cytometry*, vol. 43(2), pp. 87–93, 2001.

[13] M. Moradi and S. K. Setarehdan, *New features for automatic classification of human chromosomes: A feasibility study*, Elsevier- Pattern Recognition Letters 27, pp. 19–28, 2006.

[14] M. Popescu, P. Gader, J. Keller, C. Klein, J. Stanley and C. Caldwell, *Automatic karyotyping of metaphase cells with overlapping chromosomes*, Pergamon Computers in Biology and Medicine 29, 1999.

[15] R. J. Schalkoff, *Pattern Recognition: Statistical, Structural and Neural Approaches*, John Wiley & Sons, USA, 1992.

[16] Y. Wang and A. Dandpat, "A hybrid approach of using wavelets and fuzzy clustering for classifying multi-spectral florescence in situ hybridization images," *Int. Journal of Biomedical Imaging*, vol. 2006, pp. 1–11, 2006.

[17] Y. Wang, Q. Wu, K. R. Castleman, and Z. Xiong, "Chromosome Image Enhancement Using Multiscale Differential Operators," *IEEE Transactions On Medical Imaging*, Vol. 22, No. 5, May 2003, pp. 685–693, 2003.

[18] Q. Wu, Y. Wang, Z. Liu, T. Chen, K. R. Castleman, "The effect of image enhancement on biomedical pattern recognition," *Proceedings of the second Joint EMBS/BMES Conference Houston*, TX, USA, October 23–26, 2002, pp. 1067–1069, 2002.

6

De-Noising of Body Surface Potential Signals

Amir Hajirassouliha, Mohammad Hossein Doostmohammadi,
Mehdi Amoon, Ahmad Ayatollahi, and Ali Sadr

Iran University of Science and Technology (IUST), Iran

Body surface potential mapping (BSPM) is a non-invasive functional imaging method for reconstructing electrophysiological information about the surface of the heart. BSPM provides more diagnostic information than the 12-lead ECG. De-noising body surface potential (BSP) signals would lead to more accurate BSPM patterns therefore better diagnosis for heart problems. For de-noising of body surface potential signals we used FFT and also different types of wavelets as a de-noising tool. For the better performance evaluation of different de-noising tool's results, a quantitative value called Signal Error Ratio (SER) is used. Results are presented in different tables for comparison. After de-noising all of the BSP signals, changes in BSPM patterns are also presented to show the effects of our method in de-noising BSPs.

6.1 Introduction

Cardiac magnetic field mapping is a non-invasive method for investigating the heart's electrical activity by measuring the cardiac magnetic field, generated by intra-cardiac currents. Because electrophysiological properties such as action potential generation and conduction of cardiomyocytes are strongly dependent on cell metabolism and the intact cardiac tissue network, pathologic conditions such as myocardium infarction alter intra-cardiac electrical current flow [1].

Information about those alternations is expressed in changes in non-invasively detectable electric and magnetic field. Assessment of cardiac changes in the electrocardiogram (ECG) is a well-established diagnostic tool [1]. At least one fourth of all myocardial infarctions (MIs) are clinically unrecognized [2]. And the limitation of the conventional 12-lead ECG for optimal detection of cardiac abnormalities are widely appreciated. Body surface potential mapping (BSPM) gives even more information. BSPM is a non-invasive functional imaging method for reconstructing electrophysiological information about the surface of the heart. BSPM provides greater diagnostic information than the 12-lead ECG for MI [3, 4]. Body surface potentials (BSPs) are electric signals which are recorded from all over the torso. The record method is somehow similar to recording electrical signals in ECG. Although de-noising ECG signal is an old topic but the topic is active and still there are some new works for de-noising ECG signals [5]. De-noising BSPs is the topic that should develop more. In BSPs in addition to the shape of the signal the patterns of BSPM are also a diagnostic tool for detection heart disease so even small portion of de-noising may lead to more accurate patterns and so more accurate diagnosis. Using wavelet is one of the most powerful tools for de-noising signals. In the case of signals with the sharp part the effectiveness of wavelet has been proven. Previously wavelet has been used for de-noising of signals in ultrasound non-destructive tests (NDT) [6, 7]. Wavelet has also been used for de-noising of ECG and other kinds of biologic signals [8, 9]. In our study, we have used different types of wavelet family with both soft and hard thresholds. Besides wavelet, FFT is also known as a de-noising tool for different kind of signals [10]. We have also used FFT for de-noising of our signals. FFT has been used both by itself and also with the help of Blackman-Harris windowing. For doing a quantitative comparison between different techniques, at the first step we have used a simulated ECG signal and compute Signal Error Ratio (SER) for each de-noising tool. After collecting results from our simulated ECG we have used some of these de-noising techniques for our real BSPM data. Our BSPM data is sampled at 2 kHz and consisting of ECG data for standard 352 torso-surface sites.

6.2 Fast Fourier Transform (FFT)

The Fast Fourier Transform (FFT) is a powerful general-purpose algorithm widely used in signal analysis. FFTs are useful when the spectral information

Fig. 6.1 The entire workflow of noise reduction operation by FFT method.

of a signal is needed, such as in pitch tracking or vacating algorithms. The FFT can be combined with the Inverse Fast Fourier Transform (IFFT) in order to re-synthesize signals based on its analyses. This application of the FFT/IFFT is of particular interest in noise reduction in signals because it allows for a high degree of control of a given signal's spectral information allowing for flexible and efficient implementation of signal processing algorithms.

Traditionally the FFT/IFFT has been widely used outside of real-time for various signal analysis/re-synthesis applications. With the ability to use the FFT/IFFT in real-time, live signal processing in the spectral domain becomes possible, offering attractive alternatives to standard time-domain signal processing techniques. Some of these alternatives offer a great deal of power, run-time economy, and flexibility, as compared with standard time-domain techniques. In addition, the FFT offers both a high degree of precision in the spectral domain, and straightforward means for exploitation of this information. The entire workflow of noise reduction operation by FFT method is shown in Figure 6.1.

6.3 Wavelet

The wavelet transform is performed via a pair of filters h and g, which convolve input signal then decimate it into smooth half and detail half signals at the first level. The process then continuously operates on the smooth half until a final level reached, as shown in Figure 6.2.

The de-noising procedure proceeds in three steps:

1. Decomposition. Choose a wavelet, and choose a level N. Compute the wavelet decomposition of the signal s at level N.
2. Detail coefficients thresholding. For each level from 1 to N, select a threshold and apply soft thresholding to the detail coefficients.
3. Reconstruction. Compute wavelet reconstruction based on the original approximation coefficients of level N and the modified detail coefficients of levels from 1 to N.

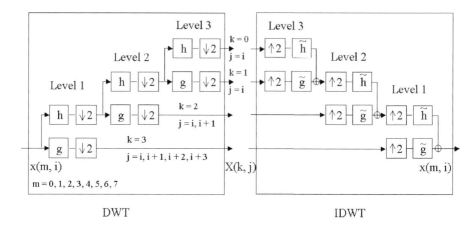

Fig. 6.2 Illustrations of DWT and IDWT in time-frequency presentation.

6.4 Simulated ECG

Simulated ECG has been used for performance evolution of our de-noising techniques with the help of Signal Error Ratio (SER). SER has been used previously for performance evolution of de-noising techniques [5]. Assume that our noisy signal $s(t)$ is a combination of a noise-free signal $x(t)$ and a noise component realization called $n(t)$. Therefore, we have this equation:

$$s(t) = x(t) + n(t), \tag{6.1}$$

After applying different types of de-noising tools, a constructed signal called $\hat{x}(t)$ is obtained. SER is written as:

$$SER = \frac{\sum_{t=0}^{L-1} x^2(t)}{\sum_{t=0}^{L-1} x(t) - \hat{x}(t)^2} \tag{6.2}$$

For our simulated ECG we have used the signal presented in Figure 6.3. The number of points is chosen to be 1639 points to well-match our real BSP signals. This simulated ECG is $x(t)$ for noise component, white Gaussian noise, n(t), is added to $x(t)$ (Figure 6.4). White Gaussian noise is added in a way that the SNR of the signal is one of these values 5 db, 10 db or 15 db. Based on equation 1, the s(t) signal is obtained and for SNR = 15 db, it is shown in Figure 6.5.

FFT has been used both simply and with the help of Blackman-Harris windowing in order to de-noise the simulated ECG. At the first step FFT of

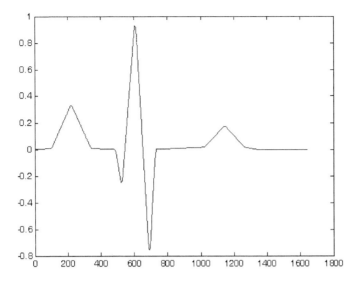

Fig. 6.3 Simulated ECG.

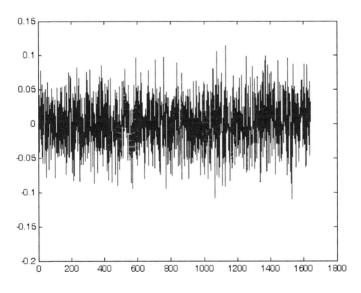

Fig. 6.4 Noise Model.

the S(t) (Equation 6.1) signal is taken. At the next step desired low-pass filter is applied in frequency domain. Then the inverse FFT is taken to reconstruct $\hat{x}(t)$. Using the low-pass filter has some limitations. If the bandwidth of the filter is selected narrow then we would lost some part of the original signal or the

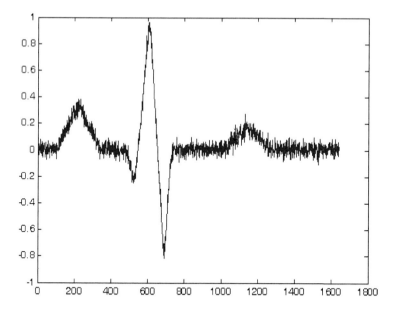

Fig. 6.5 The summation of $x(t)$ and $n(t)$ (SNR $= 15$ db).

amplitude of the original signal would be weaken. And if the bandwidth of the filter is selected wide then the de-noising process would not work properly. To overcome this problem we have used Blackman-Harris windowing and FFT in the combination. Time and Frequency domain properties of Blackman-Harris window are shown in Figure 6.6.

Equation (6.3) is used for computing the coefficients of a minimum 4-term Blackman-Harris window:

$$Wk + 1 = a_0 - a_1 \cos\left(2\pi \frac{k}{n-1}\right) + a_2 \cos\left(4\pi \frac{k}{n-1}\right)$$
$$- a_3 \cos\left(6\pi \frac{k}{n-1}\right) \qquad 0 \leq k \leq (n-1) \qquad (6.3)$$

Where the coefficients for this window are

$$a_0 = 0.35875$$
$$a_1 = 0.48829$$
$$a_2 = 0.14128$$
$$a_3 = 0.01168$$

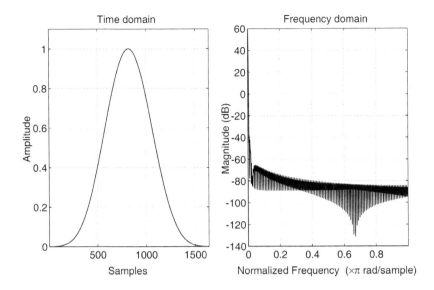

Fig. 6.6 Time domain and frequency domain properties of Blackman-Harris window.

using the Blackman-Harris window will result in some sort of averaging on noise, so, the amount of noise in low frequencies in frequency domain would decrease therefore, we can use increase the bandwidth of the filter without using some part of the original signal.

Using Blackman-Harris window has some limitations too. Because of the shape of the window in time domain (Figure 6.6) it would decrease the amplitude of the original signal. (The amplitude is less than one in all points.) To overcome this limitation after taking inverse FFT the inverse of the Blackman-Harris window is also applied to the signal.

The summery of our results for using the FFT for de-noising of our simulated data is presented in Table 6.1. The FFT is applied in different SNRs, number of samples, filter bandwidth and with or without windowing. SER is calculated in each case for performance evaluation. As it is evident in Table 6.1, using Blackman-Harris windowing would help to get better SER in similar situations and therefore, better noise reduction process. Best results are obtained when number of samples has selected to be 4096 and filter bandwidth is 150 samples. Two samples of the effects of noise reduction process on the signal presented in Figure 6.5. with the help of FFT and also FFT and windowing are shown in Figure 6.7. These signals are what we called them previously $\hat{x}t$.

Table 6.1 Summery of using FFT for de-noising of the simulated ECG.

SNR (db)	Number of samples	Filter Bandwidth (Sample)	Windowing	SER
5	2048	150	NO	13.94
5	2048	150	Blackman-Harris	14.77
5	2048	200	NO	11.60
5	2048	200	Blackman-Harris	12.22
5	4096	150	NO	16.73
5	4096	150	Blackman-Harris	18.29
10	2048	150	NO	18.29
10	2048	150	Blackman-Harris	19.12
10	4096	150	NO	22.38
10	4096	150	Blackman-Harris	21.59
10	4096	200	NO	19.79
10	4096	200	Blackman-Harris	20.74
15	2048	150	NO	23.32
15	2048	150	Blackman-Harris	23.98
15	4096	150	NO	25.85
15	4096	150	Blackman-Harris	26.90
15	4096	200	NO	24.83
15	4096	200	Blackman-Harris	25.10

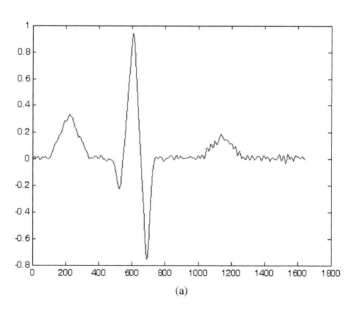

(a)

Fig. 6.7 (a) The result of using FFT for noise reduction and (b) The result of using FFT with the help of Blackman-Harris windo for the signal in Figure 6.5.

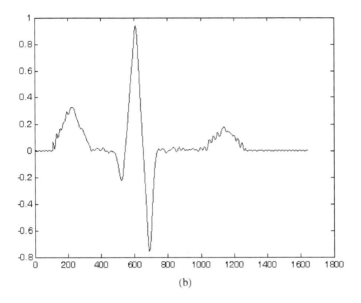

(b)

Fig. 6.7 (*Continued*)

In Figure 6.7b with the help of windowing, de-noising process has better results near the end of the signal.

6.5 Using Wavelet for De-noising of Simulated ECG

For using wavelet as a de-noising tool at the first step it is necessary to choose the type of wavelet for. Considering the shape of our signals and the shape of more common wavelet families for de-noising of ECG or BSPs we used 3 different families of wavelets consist of coiflets, daubechies and biorthogonal. The threshold of the wavelet is also important, choosing low thresholds would cause the persistence of the noise and high thresholds would result to lose the data [11]. For our study in all the cases threshold level of 5 is used for de-noising based on specifications of thresholds mentioned above. There are two kinds of thresholds, hard thresholds and soft thresholds. These thresholds are based on Equations 6.4 and 6.5.

Hard thresholding

$$x_H = \begin{cases} 0 & |x| \leq T \\ x & |x| > T \end{cases} \tag{6.4}$$

Table 6.2 SER for different types of wavelets in different SNRs.

SNR (db)	Coiflet 3		Coiflet 4		Daubechies 4	
	soft	hard	soft	hard	soft	hard
5	17.55	17.73	18.41	18.55	17.41	17.12
10	20.83	22.85	21.55	22.68	20.44	21.99
15	24.62	26.89	25.06	27.12	23.58	26.35
	Daubechies 5		Biorthogonal 2.8		Biorthogonal 4.4	
5	17.61	18.34	16.45	17.95	16.77	17.71
10	20.06	19.91	20.62	22.85	20.33	21.89
15	24.01	25.8	24.64	26.78	23.24	25.69

Soft thresholding

$$x_S = \begin{cases} 0 & |x| \leq T \\ x - T & x > T \\ x + T & x < -T \end{cases} \tag{6.5}$$

In Equations 6.4 and 6.5 T is the value which has been selected for thresholds. Hard thresholds will cause discontinues in borders of thresholds. Soft thresholds avoid this problem by decreasing the amount of the coefficients but this would lead to decrease in amplitude of the original signal, especially when the amounts of thresholds are high.

The summary of the results of using wavelet for simulated ECG in different SNRs are presented in Table 6.2. Hard thresholds have the better results in the same situations. Wavelets have better results in the case of SER in de-noising of the simulated signal, compare to FFT results. Two types of wavelet family which show better SERs are chosen. Biorthogonal 4.4 and Coiflet 4. The effects of using two types of wavelets are presented in plots in Figure 6.8.

6.6 Using FFT and Wavelets for De-noising of BSP Signals

BSP signals are consists of recorded surface potentials 352 torso-surface sites. The shapes of these signals are very similar to standard 12-lead ECG signals. We used both FFT and wavelets for de-noising of these signals. In this case because the real noise-free signals were not available SER could not be computed. The results of previous sections have been used for choosing which type of FFT or wavelet should be used. For using FFT the number of samples for filter has been chosen to be 300 samples. To show the results of de-noising process node number has been chosen. The real signal and the effects of de-noising

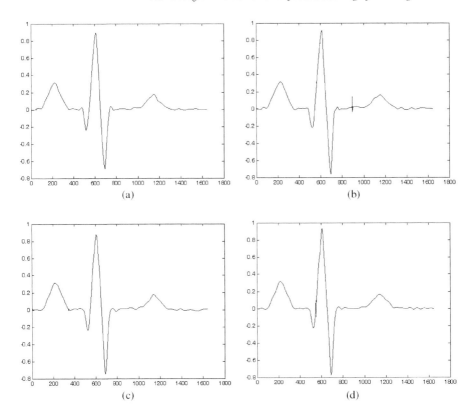

Fig. 6.8 The results of using Biorthogaonal and Coiflet wavelets for noise reduction of the signal in Figure 6.5. For each case both the hard and soft thresholds are presented. (a) Biorthogonal 4.4 (soft); (b) Biorthogonal 4.4 (hard); (c) Coiflet 4 (soft); (d) Coiflet 4 (hard).

process by using FFT have been shown in Figure 6.9. Based on our previous results presented in Table 6.1 in simulated ECG the number of samples of FFT is chosen to be 4096. The results of our real BSP signals shows that using FFT with Blackman-Harris windowing will lead to better noise-reduction process but we would lost some parts of real signal in the end of the signal.

Wavelets show better SER in simulated ECG. For BSP signals two types of wavelets which have been used for simulated ECG are used. Results are shown in Figure 6.8. Both wavelets have very good results for noise reduction. Unlike FFT with Blackman–Harris wavelets didn't change the end of the signal. Hard thresholds maintain the real shape of the signal better. Comparison of the real signal in Figure 6.10-a. and the de-noised signal in Figure 6.10-d. illustrates this fact.

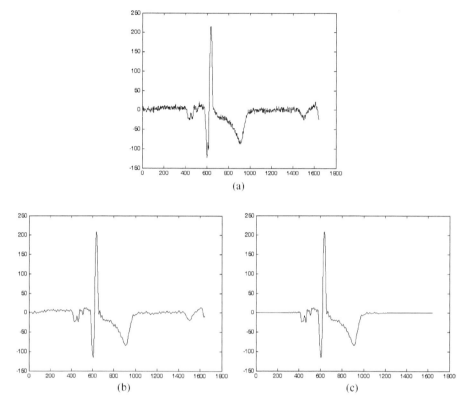

Fig. 6.9 The results of using FFT and FFT with Blackman–Harris windowing for noise reduction of the real BSP signal in figure 6.9-a. number of samples for both has been used to be 4096. (a) Real Signal; (b) FFT; (c) FFT with window.

6.7   Body Surface Potential Mapping (BSPM)

BSPM data consists of ECG data for standard 352 nodes on torso-surface. To show the effects of our de- noising on BSPM we mapped data for the QRS complex with both the original data and de-noised data. The time interval of mapping has chosen to be 75 msec. This time interval is between 550 and 700 in graphs of Figure 6.9-a. (or Figure 6.10-a.). Figure 6.11 shows the results of BSPMs. Figure 6.11-a. and Figure 6.11-c. are results for original signals and figures Figure 6.11-b. and Figure 6.11-d. are results for de-noised signals. As it is evident in BSPMs we have some changes in patterns both in front torso and back torso. The values for potentials which presented in the form of a color bar have not changed much compare two cases. This means that

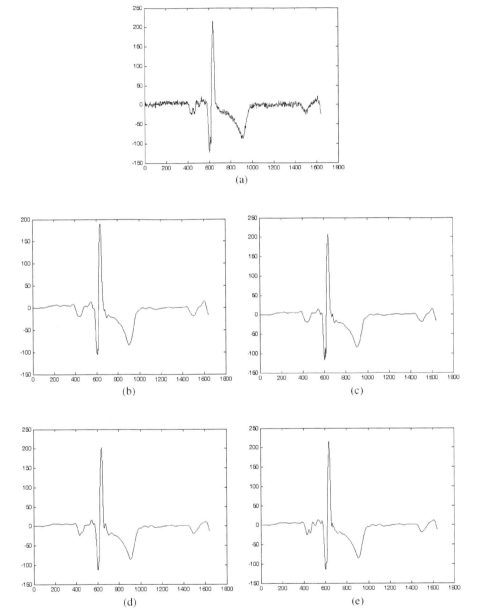

Fig. 6.10 The results of using Biorthogonl and Coiflet for noise reduction of the real BSP signal in Figure 6.10-a. Both hard and soft thresholds are used. (a) Real Signal; (b) Biorthogonal 4.4 (soft); (c) Biorthogonal 4.4 (hard); (d) Coiflet 4 (soft); (e) Coiflet 4 (hard).

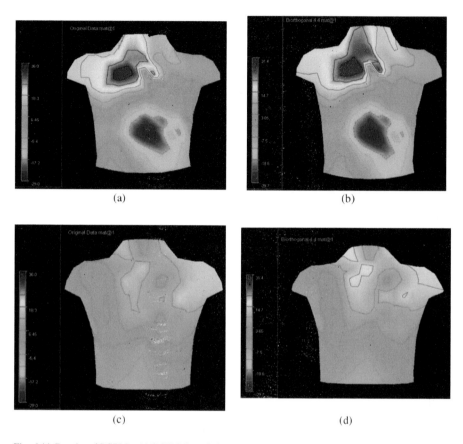

Fig. 6.11 Results of BSPMs. (a) BSPM for original data in front torso; (b) BSPM for de-noised data in front torso; (c) BSPM for original data in back torso; (d) BSPM for de-noised data in back torso.

de-noising has not affected the values of signals. In front torso the range of the changes is less than the back torso. It happens because of the strength of signals in front torso and as the consequence more SNR compare to back torso signals. Dashed lines in models are the indicator of negative values. BSPMs of de-noised data (Figure 6.11-b. and Figure 6.11-d.) have some changes in these negative lines especially near the boarders of positive and negative values.

6.8 Summary of Chapter (Conclusions)

The shape of the graphs of BSP signals shows that they are noisy. By using different wavelets from three families of wavelet and also FFT and FFT with Blackman-Harris windowing we could de-noise these signals. For better performance evolution SER parameter has been defined and simulated ECG signal has been used. The results of de-noising process for simulated ECG have been used for de-noising of real BSP signals. The results of de-noising show that although FFT has good results, wavelets have more acceptable result. Among wavelet families Biorthogonal 4.4 was our choice to show the result of de-noising BSPs. After de-noising, BSPs are mapped on standard torso models. The changes in the shape of the patterns were obvious in both front and back torso. Because of the importance of these patterns in detection of different kinds of cardiovascular disease de-noising of BSPs are completely necessary. Using wavelet for de-noising is a good choice and between all kinds of wavelets our result show that biorthogonal 4.4 will lead to better results. The de-noised signals have changed the patterns in BSPMs. These changes were clear in both front and back torso models.

Problems

Problem 6.1: What are the advantages of using body surface potentials instead of standard 12-lead ECG?

Problem 6.2: Please name the important factors for choosing a de-noising method for an effective de-noising for a desire signal.

Problem 6.3: When it is better to choose FFT with windowing for de-noising of a desire signal?

Problem 6.4: How the number of FFT points is selected?

Problem 6.5: Assume that we had a signal which its amplitude contains valuable information, which kind of de-noising method do you suggest and why?

References

[1] Thaler, Malcolm S., The Only EKG Book You'll Ever Need, The, 5th Edition, Lippincott Williams & Wilkins, pp. 10–59, 2007.
[2] S. Sheifer, T. Manolio, and B. Gresh, "Unrecognized myocardial infarction". *Ann Intern Med*, pp. 135–801, 2001.
[3] S.J. Maynard, B. Taccardi, and P.R. Ershler, "Body surface mapping improves early diagnosis of acute myocardial imfarctions in patients with chest pain and left bundle branch block", *Heart Journal*, 89, pp. 998–1002, 2003.

[4] A. Medvegy, I. Preda, and P. Savard, "New body surface isopotential map evaluation method to detect minor potential losses in non-Q-wave myocardial infarction", *Circulation*, pp. 1101–1115, 2000.

[5] M. Blanco-Velasco, B. Weng, and K.E. Barner, "ECG signal denoising and baseline wander correction based on the empirical mode decomposition", Computers in Biology and Medicine, Volume 38, Issue 1, pp. 1–13, January 2008.

[6] J.C. Lazaro, "Noise reduction in ultrasonic NDT using discrete wavelet transform", *Processing IEEE Ultrasonics Symposium*, Volume 1, October 2002, pp. 777–780.

[7] F. Ykhlef, M. Arezki, A. Guessoum, and D.A. Berkani, "wavelet denoising method to improve detection with ultrasonic signal", *IEEE ICIT '04 IEEE International Conference*, Volume 3, December 2004, pp. 1422–1425.

[8] H. Tirtom, M. Engin, and E.Z. Engin, "Enhancement of time-frequency properties of ECG for detecting micropotentials by wavelet transform based method", *Expert Systems with Applications*, Volume 34, Issue 1, pp. 746–753, January 2008.

[9] B. Molavi and A. Sadr, "Optimum wavelet design for noise reduction and feature extraction", *Communications and Information Technologies, ISCIT '07*, October 2007, pp. 1096–1101.

[10] S.V. Vaseghi, *Advanced Digital Signal Processing and Noise Reduction*, 2nd Edition, John Wiley & Sons, 2000.

[11] D.L. Donoho, "De-noising by soft-thresholding", *IEEE Transactions on Information Theory*, Volume 41, Issue 3, pp. 613–627, May 1995.

7

Single Channel Wireless EEG: Proposed Application in Train Drivers

Budi Yap, Sara Lal and Peter Fischer

University of Technology, Sydney, Australia

This chapter describes the development of wireless electroencephalogram (EEG) to be used in a fatigue countermeasure device for train drivers. EEG can be used as an indicator of fatigue. Several studies have shown that slow wave brain activities, delta (0–4 Hz) and theta (4–8 Hz), increase as an individual becomes fatigued, while the fast brain activities, alpha (8–13 Hz) and beta (13–35 Hz), decrease. However, EEG measurement is via a complex piece of equipment that is generally used in laboratory-based studies. In order to develop an EEG based fatigue countermeasure device, there is a need for a simple and wireless EEG monitor.

7.1 Introduction

In the late 1800s Richard Caton (1842–1926) first reported the presence of biopotentials on the surface of the human skull [1]. The electroencephalography (EEG), pioneered by Hans Berger, measures the electric potentials on the scalp and provides continuous measure of cortical activity [2, 3]. EEG has played an important role due to its non-invasiveness and the capability of long term measurement in the field of epilepsy, sleeping disorder, fatigue, and various neurological conditions [2, 3, 4, 5].

An example of EEG application is in the detection of driver fatigue. Driver fatigue is a process that involves successive episodes of micro-sleeps, where the subject may go in and out of a fatigue state [6]. EEG has been used in many driver studies conducted in the lab and field [5, 7, 8, 9, 10], and is one of the most reliable indicators of fatigue [11]. Studies have found that EEG has acceptable test and retest reliability [6, 12, 13, 14, 15]. Subsequently, Lal & Craig [16] proposed that EEG could be used as an indicator of fatigue and proposed driver fatigue algorithms. Later, Jap et al. [17] proposed an EEG-based fatigue algorithm for application in the train driving environment. Lal & Craig [7] reported increases in slow wave brain activity with driver fatigue. Delta (0–4 Hz) and theta (4–8 Hz) activities have been found to increase as one gets fatigued [16]. Jap et al. [17] found significant decreasing trends of alpha (8–13 Hz), and beta (13–35 Hz) activities as fatigue increases. Beta activity has also been linked to task performance. As beta level decreases, task performance has also been found to decrease [5].

However, the conventional EEG machine involves attaching the machine to a computer through a data cable, and this introduces some limitations in the applications of EEG monitoring as a fatigue detector [4]. In the actual driving environment, recording EEG data using a conventional EEG machine may not be feasible, because EEG is a complicated piece of technology for laboratory environment setup. The cables from the EEG machine may be intrusive and hinder the driver. Therefore, a simple, portable, and wireless EEG is required for application in fatigue monitoring.

This technical chapter proposes the design of a single channel wireless EEG device that is suitable for fatigue detection in train drivers. The EEG measurement was designed for bipolar montage recording. The recording in bipolar montage is obtained from the difference between two active electrodes that are adjacent to each other on a particular brain site, e.g. frontal site [18]. The EEG electrodes are aimed to be placed inside a headband at either the frontal or temporal sites, according to the 10–20 electrode placement standard, in order to achieve uniformity in the interpretation of recorded results [19, 20]. Refer to Figure 7.1 for the 10–20 electrode placement system.

The following sections will describe the development of a single channel wireless EEG prototype, and will present a few results on EEG recordings obtained from the prototype.

7.2 Hardware Development of Wireless EEG

The amplitude of the EEG signal depends on how synchronous the activities of the underlying neurons are [21]. The human EEG signal ranges in frequency

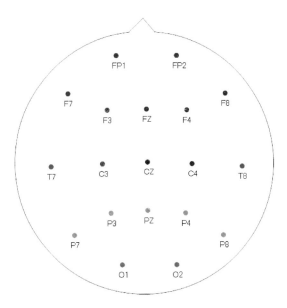

Fig. 7.1 The international 10–20 [19] system seen from above the head; FP = Fronto-Polar, F = frontal, C = central, T = temporal, P = parietal, O = occipital, Z = midline central electrode.

from 0.5–100 Hz, with amplitudes of 1–300 μV measured at the surface of the skull [3, 20, 22]. These signal characteristics have inherent challenges to their measurement. Firstly, 50 Hz noise is present within the EEG frequency range. Significant electrical noise is present at this frequency in most environments. The very low frequency EEG signal makes it susceptible to 50 Hz noise. The signal may further be corrupted by impedance imbalance at the skin-electrode interface, electrode half-cell potential, and movement artifacts [3].

A prototype wireless EEG system has been developed in order to extract EEG signal from human subjects, and to be analysed for fatigue detection. The hardware module consists of transmitter and receiver ends. While the receiver end is digital, the transmitter end consists of analog and digital sections. A block diagram of the system is depicted in Figure 7.2. The prototype is designed to operate with a 6-volt (V) battery. A voltage divider and voltage follower are used to obtain a $+/-3$ V supply.

7.2.1 Electrodes

The electrodes used for this system is the MLA2503 — Shielded Lead Wires Snap-On electrodes (refer to Figure 7.3). The lead wires are 98 cm in length

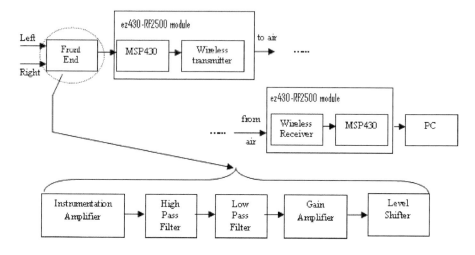

Fig. 7.2 Block diagram of the EEG system prototype.

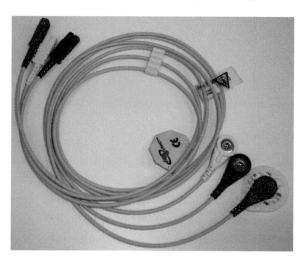

Fig. 7.3 Shielded Lead Wire Electrodes (ADInstruments, MLA2503 Shilded Lead Wires, Australia).

with 4 mm snap-on connectors used with 3M Red Dot Ag/AgCl (silver/silver-chloride) monitoring electrode. An adequate amount of conductive gel is applied on the electrodes prior to attaching them to the subject.

One of the advantages of using the Ag/AgCl electrodes is the superior performance in term of low-frequency noises [23]. However, from the equivalent circuit of a pair of electrodes in Figure 7.4, it is obvious that the balance

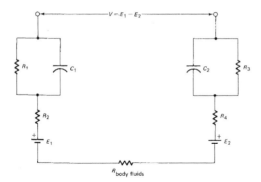

Fig. 7.4 The equivalent circuit of 2-electrode biopotential measurement [24].

between left and right electrodes will eliminate undesirable over-potential across the electrodes.

7.2.2 Analog Front End

The subject is connected to the front end of the transmitter through three EEG electrodes. The front end is mainly the analogue electronics, which consists of an instrumentation amplifier followed by signal amplification and condition-ing stages (i.e. direct current (DC) restorator, gain amplifier, high-pass and low-pass filters, variable attenuation and signal level shifter). The front-end schematic is shown in Figure 7.5.

7.2.3 AD620 Instrumentation Amplifier

The instrumentation amplifier is used to amplify the difference of the signal from the left and right EEG electrodes, and simultaneously reject common mode noise at both inputs. The AD620 instrumentation amplifier (Analog Devices, USA) was the instrumentation amplifier chosen for this design as it offers low power and high Common Mode Rejection Ratio (CMRR). Without a high CMRR, common mode voltages, which are typically 1V [25], will get amplified and prevent the EEG signal from being recovered. The CMRR ratio of AD620 is in the range of 90 to 110 dB for Gain $G = 5$ in the frequency range from DC to 60 Hz (refer to Figure 7.6) [26]. Another reason to choose AD620 as the preamplifier stage is due to its low voltage noise (9 nV/$\sqrt{\text{Hz}}$) and low input current noise (0.1 pA/$\sqrt{\text{Hz}}$).

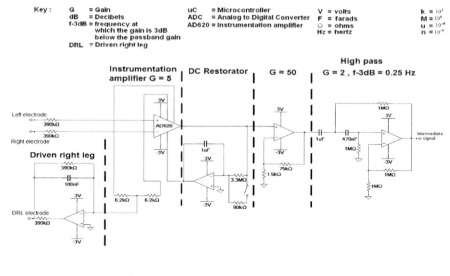

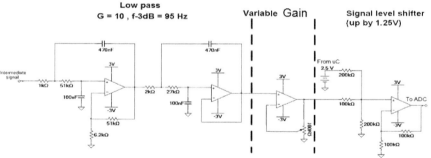

Fig. 7.5 Schematics of the Analog Front End.

In the subsequent stages, LMC6484 (low-power Operational Amplifier (op-amp), National Semiconductor, USA) Integrated Circuits (ICs) are used to implement all signal conditioning circuitry. The LMC6484 has low-power consumption making it suitable for portable battery powered applications.

7.2.4   DC Restorator

The DC restorator circuit is used to eliminate DC offset that would otherwise saturate the op-amps subsequent to the instrumentation amplifier. The DC restorator is implemented by using an op-amp in the feedback loop of the AD620. The DC restorator has two settings, namely the 'monitor', and the

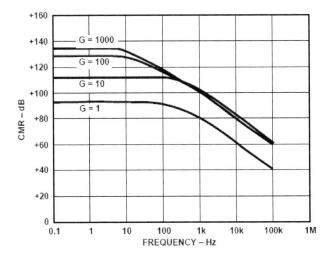

Fig. 7.6 Common Mode Rejection Ratio (CMRR) of AD620 instrumentation amplifier [26]; Hz = Hertz; dB = decibel; G = gain.

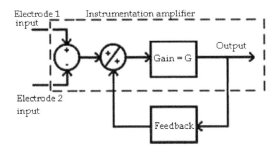

Fig. 7.7 DC Restorator Model.

'quick restore'. The 'quick restore' setting can be used to speed up the recovery of the output if it becomes saturated [3]. The model in Figure 7.7 was used to design the DC restorator.

7.2.5 High Pass and Low Pass Filters

The high-pass and low-pass filters are used to cut-off the signal with frequency components below 0.25 Hz and above 95 Hz. The high pass filter is implemented by using a 2nd-order Sallen-Key filter [27]. It attempts to eliminate the DC potential present at the electrode. The low pass filter is implemented by using two 2nd-order Sallen-Key filter stages. Therefore it is effectively a

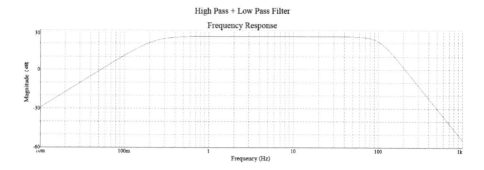

Fig. 7.8 Frequency response of high pass and low pass filters (dB = decibel; Hz = Hertz).

4th-order filter. Refer to Figure 7.8 for the frequency response of the high pass and low pass filters.

7.2.6   Driven Right Leg

The Driven Right Leg (DRL) is incorporated in the design to reduce common-mode noise, such as 50 Hz interference unavoidably coupled into the subject and electrodes. The name, Driven Right Leg, has been preserved from its first use in electrocardiography (ECG) equipment [28], although this electrode is not connected to the subject's right leg for EEG recordings. As shown in Figuire 7.7, the DRL is a feedback circuit similar to the DC restorator. It feeds the inverse of the common-mode voltage back to the human subject, which acts to reduce the common-mode noise present at the active electrodes [28].

7.2.7   MSP430 and Wireless Modules

The first prototype of the digital section utilises the eZ430-RF2500 development tool (Texas Instruments, USA). It contains both the MSP430 microcontroller and CC2500 wireless module (see Figure 7.2). The microcontroller is suitable for low power operation as it only consumes 0.7 microAmps (uA) while in stand-by mode. In addition, it processes 16-million instructions per second, and is equipped with two 10-bit Analog-to-Digital (A-D) converters with maximum sampling rates of 200 Kilo samples per second.

The communication protocol running on CC2500 is SimpliciTI, which is a dedicated low data rate protocol for low power applications. The wireless transmission utilises the 2.4 Giga Hertz (GHz) Industrial, Scientific and Medical (ISM) frequency [29].

At the transmitter end, the microcontroller takes the output of the analog front end and performs the A-D conversion. A fatigue detection algorithm could be implemented in the microcontroller at the transmitter end to detect fatigue before the digitised signal is transmitted to the CC2500 module of the receiver end.

At the receiver end, the same eZ430-RF2500 development board, which is used at the transmitter end, is connected to a PC via a USB slot.

7.3 Experiments and Results

Figure 7.9 shows the prototype that has been developed for the project. The EEG signal was transmitted from the TI MSP430 Wireless transmitter to the computer. To aid the experiment, a testing program was built (using LabView 8.2, National Instruments, USA) in order to display the received signal at the receiver end. The development prototype was tested on a human subject.

Firstly, a 10 Hz, 100 mV sine-wave was used to verify the accuracy of the Wireless EEG prototype. The amplitude of the sine wave was further scaled down by means of voltage divider into a few hundred microvolts before being pumped into the preamplifier stage. Figure 7.10 displays the received sine wave pattern at the receiver's end, which proves the accuracy of the Wireless EEG prototype.

The development prototype was then tested on a human subject to capture ECG signal (heart beat) by connecting 3 electrodes from the device to the left wrist, right wrist and right leg of the subject. The subject was asked to remain seated and stop moving in order to avoid movement artifacts. With the aid of the testing program, the device was able to successfully capture the PQRS complexes of the ECG signal from the subject, as shown in Figure 7.11.

Finally, the circuit gain was adjusted to allow sufficient amplification for acquiring a typical EEG signal. The prototype was then used to capture the EEG signal from electrodes placed on the frontal lobe of the head. This was performed by connecting a pair of shielded electrodes to the pre-frontal brain sites (FP1 and FP2) and the DRL electrode was connected to the ear lobe of the subject. The subject was again asked to remain seated and stay still to avoid movement artifacts. Figures 7.12 and 7.13 depict the EEG signal

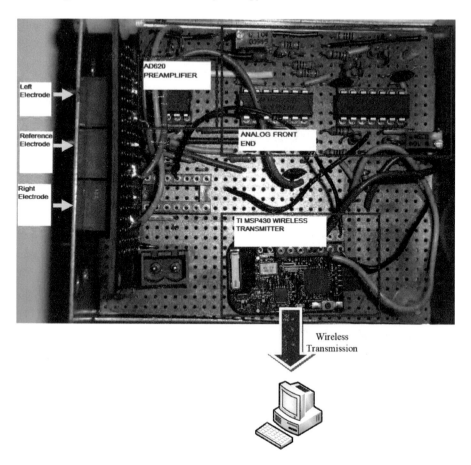

Fig. 7.9 Prototype of Wireless EEG.

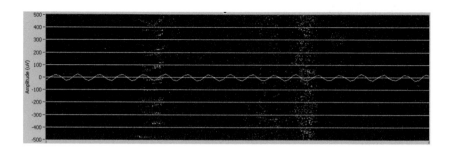

Fig. 7.10 Display of the received sine wave from the wireless transmission.

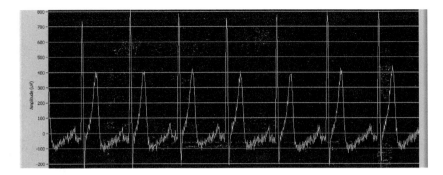

Fig. 7.11 ECG reading from the EEG prototype.

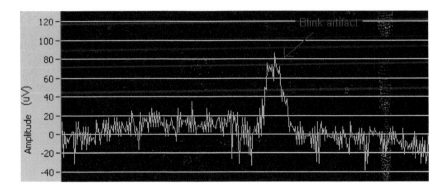

Fig. 7.12 EEG reading from FP1 and FP2 with 'eyes opened' — blink artifact.

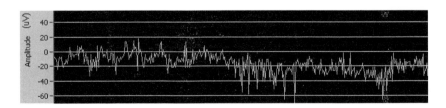

Fig. 7.13 EEG reading from FP1 and FP2 with eyes closed.

during eyes-open and eye-closure, respectively. In Figure 7.12, it was noted that an eye blink would cause an artifact in the EEG waveform. In contrast, the EEG reading with the eyes closed in Figure 7.13 showed fewer artifacts.

7.4 Discussion and Future Work

The area of safety in transportation has received attention due to a high number of accidents caused by driver fatigue [30]. Railway is a relatively safe transportation means. The number of train accidents and injuries in Australia is relatively low, when compared to road and aviation industries [31]. However, given the fact that the travelling speed of a train is relatively faster than a car, a train carriage is much heavier than a car, and that trains' braking time is slower than a car (about 1.1 m/s^2) [32], the impact that it creates in an accident causes more injuries due to a usually large numbers of passengers involved, destruction to the surrounding area and emotional and financial costs to community. Unlike the driver of a car who can swerve his or her vehicle to avoid collisions, a train driver can only apply the emergency brake, sound the horn and hope that the train will stop in time before the collision [33, 34], and if the driver is in a fatigue state even these safety measures may fail.

In the train driving environment, human errors caused by fatigue contributes to about 75% of accidents [35, 36], and research has shown that driving while fatigued is just as dangerous as driving with a blood alcohol concentration of 0.05 - 0.1% [37]. Train drivers are required to work in a 24-hour irregular and rotating shifts, including weekends [38], and irregular shifts are related to increased fatigue level and have been correlated to higher risk of fatigue-related accidents [39]. Therefore, there is a need for an automated and non-intrusive fatigue countermeasure device for drivers in this environment.

Electroencephalography has been shown to be a reliable fatigue indicator [11]. Several studies have already investigated the efficacy of EEG as a fatigue indicator. Lal and Craig [7], Lal et al. [16], Jap et al. [40] and Eoh et al. [5] proposed fatigue algorithm derived from the four EEG frequency bands. Tietze [41] explored detection of fatigue using the occurrence of 'alpha-spindles', which consisted of short periods of high amplitude alpha activity, and could be visually detected. Others have proposed the use of neural network, wavelet transform, or Independent Component Analysis (ICA) algorithm to detect fatigue [42, 43, 44, 45].

However, the complexity of an EEG system makes it difficult to deploy it as a fatigue countermeasure device in the actual train driving environment [46]. Simple and wireless EEG electrodes that can be worn by drivers while driving need to be developed, before fatigue countermeasures using EEG will become viable. These simple EEG measurement systems will also need to be able to reduce signal noise that is received from surrounding environment. Since drivers are also dynamic human beings, they may use mobile phones, which introduce high-frequency noise into the EEG recording. The noise may affect the recordings and reduce the accuracy of the device to detect fatigue.

The current chapter proposes the prototype development of a simple single-channel EEG device that transmits wirelessly to a computer. Bipolar montage has been used for the analog front-end, by amplifying the difference of the EEG signals from the left and the right active electrodes, and the DRL electrode has been implemented to reduce the common-mode noise that are present at the active electrodes. The eZ430-RF2500 development tool has been utilized to transmit EEG signals wirelessly.

The simulation result of the device shows that noise is present when recording ECG or EEG (Figures 7.11 to 7.13). Electrical noise, such as 50 Hz interference, has been filtered by the high- and low-pass filters implemented in the design. The noise that is captured in the recording may be due to insufficient contact being established between the electrodes and the skin, or subject's involuntary movements.

For the current prototype development of the wireless EEG, the device transmitted the sampled EEG signal from the analog front end to the computer. The computer then processed the received EEG signal, transformed it to frequency domain using Fast Fourier Transform (FFT) computation, and detected whether or not the person is in a fatigued state. The FFT computation may be able to be incorporated into the microcontroller. However, the on-board memory of the MSP430 microcontroller was considerably small to be able to perform effective Fast Fourier Transform (FFT) computation. Future work may incorporate external memory into the microcontroller design to address the current issue. Fatigue detection may also be performed within the wireless EEG device if the device has enough memory capacity, and fast computation speed.

The current prototype utilized the eZ430-RF2500 development tool from Texas Instruments, USA, since the on-board microcontroller was also acting as a transceiver. Other wireless technologies, such as Bluetooth or ZigBee, may also need to be explored to discover a better quality device. Bluetooth is a non-expensive device with short-range data communication, low data rate and low power consumption [47, 48]. Data can be transmitted to a distance of up to 10 m with low power usage of less than 1 Milliwatt (mW) [49, 50]. ZigBee protocol is similar to Bluetooth protocol with some minor differences in functionalities, because this protocol was developed to provide improvements to Bluetooth technology [51]. However, when compared with Bluetooth, some advantages and disadvantages still exist in ZigBee technology. The coverage area for ZigBee is comparatively larger with a radius of 70 m, with 64,000 nodes that can be connected to a master device, compared to only 7 slaves for Bluetooth devices [51]. However, the data rate is reduced to approximately 250 Kbps, and ZigBee is only suitable for small data packet transmissions, while Bluetooth is capable to transmit large data sets, audio, graphics, or files [51]. Other types of wireless technologies may also be explored in the future development.

The current wireless EEG device has to be miniaturized in the future work in order to fit into a headband or a cap, as shown in Figure 7.14. Such design

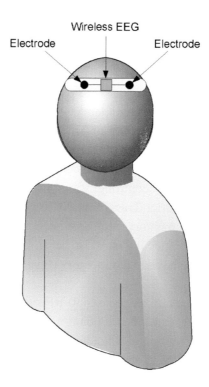

Fig. 7.14 Proposed wireless EEG headband for the development of a fatigue countermeasure device.

implementation will be suitable for train driving environment, since the wireless EEG will be non-intrusive in a driving situation.

Fatigue detection is one of the areas that wireless EEG can be utilized. Applications of wireless EEG can be expanded to other areas, such as simple monitoring of the state of an individual in the daily life environment outside laboratories or hospitals. However, for every development of wireless EEGs, the design of the device has to conform with the safety requirements according to the IEC 60601-1:1988 (safety standard of medical electrical equipment). The next stage of the prototype development will need to address this safety issue.

7.5 Acknowledgement

The research was supported by an ARC Linkage grant Australia (LP0560886) and by SENSATION Integrated Project (FP6-507231) co-funded by the Sixth Framework Programme of the European Commission under the Information Society Technologies priority.

References

[1] L. F. Haas, "Hans Berger (1873-1941), Richard Caton (1842-1926), and electroencephalography. (Neurological Stamp)," *Journal of Neurology, Neurosurgery and Psychiatry*, vol. 74, p. 9, 2003.

[2] W. De Clercq, B. Vanrumste, J.-M. Papy, W. Van Paesschen, and S. Van Huffel, "Modeling Common Dynamics in Multichannel Signals With Applications to Artifact and Background Removal in EEG Recordings," *IEEE Transactions on Bioimedical Engineering*, vol. 52, pp. 2006–2015, 2005.

[3] J. J. Carr and J. M. Brown, *Introduction to Biomedical Equipment Technology*, 4th ed. New Jersey, USA: Prentice Hall, 2001.

[4] R. Lin, R.-G. Lee, C.-L. Tseng, Y.-F. Wu, and J.-A. Jiang, "Design and Implementation of Wireless Multi-channel EEG Recording System and Study of EEG Clustering Method," *Biomedical Engineering - Applications, Basis & Communications*, vol. 18, pp. 276–283, 2006.

[5] H. J. Eoh, M. K. Chung, and S.-H. Kim, "Electroencephalographic study of drowsiness in simulated driving with sleep deprivation," *International Journal of Industrial Ergonomics*, vol. 35, pp. 307–320, 2005.

[6] S. K. L. Lal and A. Craig, "Reproducibility of the spectral components of the electroencephalogram during driver fatigue," *International Journal of Psychophysiology*, vol. 55, pp. 137–143, 2005.

[7] S. K. L. Lal and A. Craig, "Driver Fatigue: Electroencephalography and Psychological Assessment," *Psychophysiology*, vol. 39, pp. 313–321, 2002.

[8] T. Åkerstedt, G. Kecklund, and A. Knutsson, "Manifest Sleepiness and the Spectral Content of the EEG during Shift Work," *Sleep*, vol. 14, pp. 221–225, 1991.

[9] G. Kecklund and T. Åkerstedt, "Sleepiness in long distance truck driving: an ambulatory EEG study of night driving," *Ergonomics*, vol. 36, pp. 1007–1017, 1993.

[10] O. G. Okogbaa, R. L. Shell, and D. Filipusic, "On the investigation of the neurophysiological correlates of knowledge worker mental fatigue using the EEG signal," *Applied Ergonomics*, vol. 25, pp. 355–365, 1994.

[11] P. Artaud, S. Planque, C. Lavergne, H. Cara, P. de Lepine, C. Tarriere, and B. Gueguen, "An on-board system for detecting lapses of alertness in car driving," in *14th E.S.V. Conference, Session 2 — Intelligent Vehicle Highway System and Human Factors*, Munich, Germany, 1994.

[12] A. J. Tomarken, R. J. Davidson, R. E. Wheeler, and L. Kinney, "Psychometric properties of resting anterior EEG asymmetry: temporal stability and internal consistency," *Psychophysiology*, vol. 29, pp. 576–592, 1992.

[13] V. E. Pollock, L. S. Schneider, and S. A. Lyness, "Reliability of topographic quantitative EEG amplitude in healthy late-middle-aged and elderly subjects," *Electroencephalography and Clinical Neurophysiology*, vol. 79, pp. 20–26, 2002.

[14] T. Gasser, P. Bacher, and H. Steinberg, "Test-retest reliability of spectral parameters of the EEG," *Electroencephalography and Clinical Neurophysiology*, vol. 60, pp. 312–319, 1985.

[15] T. Gasser, L. Sroka, and J. Möcks, "The transfer of EOG activity into the EEG for eyes open and closed," *Electroencephalography and Clinical Neurophysiology*, vol. in press, 1985.

[16] S. K. L. Lal, A. Craig, P. Boord, L. Kirkup, and H. Nguyen, "Development of an algorithm for an EEG-based driver fatigue countermeasure," *Journal of Safety Research*, vol. 34, pp. 321–328, 2003.

[17] B. T. Jap, S. Lal, P. Fischer, and E. Bekiaris, "Using Spectral Analysis to Extract Frequency Components from Electroencephalography: Application for Fatigue Countermeasure in Train Drivers," in *The 2nd International Conference on Wireless Broadband and Ultra Wideband Communications (AusWireless 2007)* Sydney, Australia, 2007.

[18] J. M. Stern and J. Engel, *Altas of EEG Patterns*. USA: Lippincott Williams & Wilkins, 2005.

[19] H. H. Jasper, "Report of the committee on methods of clinical examination in electroencephalography : 1957," *Electroencephalography and Clinical Neurophysiology*, vol. 10, pp. 370–375, 1958.

[20] A. J. Rowan and E. Tolunsky, *Primer of EEG: With a Mini-Atlas*. USA: Elsevier Science, 2003.

[21] M. F. Bear, B. W. Connors, and M. A. Paradiso, "Rhythms of the Brain," in *Neuroscience: exploring the brain*, 2nd ed, M. F. Bear, B. W. Connors, and M. A. Paradiso, Eds. USA: Lippincott Williams & Wilkins, 2001, pp. 606–636.

[22] J. Malmivuo and R. Plonsey, *Bioelectromagnetism: Principles and Applications of Bioelectric and Biomagnetic Fields*: Oxford University Press, USA, 1995.

[23] M. R. Neuman, "Biopotential Electrodes," in *Medical Instrumentation: Applicantion and Design*, 3rd ed, J. G. Webster, Ed. USA: John Wiley and Sons, 1998.

[24] L. Cromwell, Weibell, Fred J and Pfeiffer, Erich A, *Biomedical Instrumentation and Measurements*, 2nd ed. New Jersey: Prentice-Hall, 1980.

[25] B. B. Winter and J. G. Webster, "Reduction of Interference Due to Common Mode Voltage in Biopotential Amplifiers," *IEEE Transactions on Bioimedical Engineering*, vol. 30, pp. 58–62, 1983.

[26] Analog Devices, "Low Cost, Low Power Instrumentation Amplifier — AD620," Analog Devices, 1999.

[27] D. J. Comer and D. T. Comer, *Advanced Electronic Circuit Design*. Great Britain: Wiley, 2003.

[28] B. B. Winter and J. G. Webster, "Driven-Right-Leg Circuit Design," *IEEE Transactions on Bioimedical Engineering*, vol. 30, pp. 62–66, 1983.

[29] Texas Instrument, "SimpliciTI Network Protocol," viewed 07 May 2008, <http://focus.ti.com/docs/toolsw/folders/print/simpliciti.html>

[30] D. F. Dinges, "An overview of sleepiness and accidents," *Journal of Sleep Research*, vol. 4, pp. 4–14, 1995.

[31] National Transport Commission (NTC), "Fatigue Management Within The Rail Industry: Review of Regulatory Approach," Melbourne, Australia 2004.

[32] M. Ashiya, S. Sone, Y. Sato, and A. Kaga, "Application of pure electric braking system to electric railcars," in *Proceedings of 6th International Workshop on Advanced Motion Control*, 2000, pp. 163–168.

[33] A. Austin and P. D. Drummond, "Work problems associated with suburban train driving," *Applied Ergonomics*, vol. 17, pp. 111–116, 1986.

[34] A. Jabez, "Back in control," *Nursing Times*, vol. 89, pp. 46–47, 1993.

[35] G. J. S. Wilde and J. F. Stinson, "The monitoring of vigilance in locomotive engineers," *Accident Analysis & Prevention*, vol. 15, pp. 87–93, 1983.

[36] G. D. Edkins and C. M. Pollock, "The influence of sustained attention on Railway accidents," *Accident Analysis & Prevention*, vol. 29, pp. 533–539, 1997.

[37] G. D. Roach, J. Dorrian, A. Fletcher, and D. Dawson, "Comparing the effects of fatigue and alcohol consumption on locomotive engineers' performance in a rail simulator," *Journal of Human Ergology*, vol. 30, pp. 125–130, 2001.

[38] RailCorp, "Careers in Service Delivery," viewed 26 March 2008, <http://www.railcorp.info/careers/careers_in_service_delivery>

[39] T. Åkerstedt and P. M. Nilsson, "Sleep as restitution: an introduction," *Journal of Internal Medicine*, vol. 254, pp. 6–12, 2003.

[40] B. T. Jap, S. Lal, P. Fischer, and E. Bekiaris, "Using EEG spectral components to assess algorithms for detecting fatigue," *Expert Systems with Applications*, vol. 36, pp. 2352–2359, 2009.

[41] H. Tietze, "Stages of wakefulness during long duration driving reflected in alpha related events in the EEG," in *Proceedings of the International Conference on Traffic and Transport Psychology ICTTP*, Bern, Zwitserland, 2000.

[42] T.-P. Jung, S. Makeig, M. Stensmo, and T. J. Sejnowski, "Estimating alertness from the EEG power spectrum," *IEEE Transactions on Bioimedical Engineering*, vol. 44, pp. 60–69, 1997.

[43] A. Subasi, "Automatic recognition of alertness level from EEG by using neural network and wavelet coefficients," *Expert Systems with Applications*, vol. 28, pp. 701–711, 2005.

[44] M. K. Kiymik, M. Akin, and A. Subasi, "Automatic recognition of alertness level by using wavelet transform and artificial neural network," *Journal of Neuroscience Methods*, vol. 139, pp. 231–240, 2004.

[45] C.-T. Lin, R.-C. Wu, S.-F. Liang, W.-H. Chao, Y.-J. Chen, and T.-P. Jung, "EEG-based drowsiness estimation for safety driving using independent component analysis," *IEEE Transactions on Circuits and Systems I: Regular Papers*, vol. 52, pp. 2726–2738, 2005.

[46] A. Williamson and T. Chamberlain, "Review of on-road driver fatigue monitoring devices," New South Wales, Australia: NSW Injury Risk Management Research Centre, University of New South Wales, 2005.

[47] J. Yao, R. Schmitz, and S. Warren, "A Wearable Standards-Based Point-of-Care System for Home Use," in *international Conference of the IEEE EMBS*, Cancun, Mexico, 2003.

[48] C.-H. Shih, K. Wang, and H.-C. Shih, "An adaptive bluetooth packet selection and scheduling scheme in interference environments," *Computer Communications*, vol. 29, pp. 2084–2095, 2006.

[49] H. A. Thompson, "Wireless and Internet communications technologies for monitoring and control," *Control Engineering Practice*, vol. 12, pp. 781–791, 2004.

[50] I. Chlamtac, M. Conti, and J. J. N. Liu, "Mobile ad hoc networking: imperatives and challenges," *Ad Hoc Networks*, vol. 1, pp. 13–64, 2003.

[51] N. Wang, N. Zhang, and M. Wang, "Wireless sensors in agriculture and food industry–Recent development and future perspective," *Computers and Electronics in Agriculture*, vol. 50, pp. 1–14, 2006.

8

Algorithm of Remote Monitoring ECG Using Mobile Phone: Conception and Implementation

Rachid Merzougui and Mohammed Feham

Faculty of Engineering Science of Tlemcen STIC Laboratory Chetouane Tlemcen Algeria, Algeria

8.1 Introduction

Telemedicine appears to be a medical reality: it has already imposed the use of portable units as mobile phones. The recent technological advances of mobile telecommunications networks applied to the medical domain (medical imaging, debit of transmission, confidentiality of data, the conviviality of systems, etc.) and the miniaturization of devices open perspectives for medical development of remote monitoring in term of acroissement of the efficiency and the care's quality, the knowledge's sharing or reduction of public health cost. These new technologies have led to the emergence of a wide variety of new ways for users to access and use information which interests it anywhere and anytime [1]. Then, today a simple mobile phone can contribute effectively to safeguard of the human lives.

This paper describes an application using mobile wireless networks to treat and monitor the state of patient and aged ones at home.

The suggested solution is an implement of an algorithm which transmits the data of the patient via a wireless communication in the purpose to exploit a mobile phone for medical monitoring (detection, calculation of cardiac frequency, visualization of ECG signal on the screen of the mobile phone...).

The majority of the works undertaken in this field carry out the analysis of the signals on large server (great capacities, better resolutions…). We suggest in this article to introduce the complete analysis of ECG signal for the cardiac persons on a simple mobile phone by respecting its necessary constraints.

Thus the orientation of our works in this sense was dictated by mobile networks services, simplicity of management and adaptation to the context of mobility with a low cost of exploitation.

In following sections, we present different subjects concerning the solution of detecting critical situations on limited resources by an adaptation strategy to design and develop health services.

Indeed, it is necessary at first to formulate precisely the problem to identify the areas of research which effectively require to be addressed.

8.2 Problem

In the context of this paper, we are particularly focused on the carried out operations in the processing and analysis of the received medical signals from the installed sensors at home. This step is fundamental to an effective exploitation of the potentialities for collecting a big amount of data which improves monitoring to ensure a permanent safety of patients at home and prevents a degradation of their Health state. The extracted information on the patient's situation must be relevant for help to diagnosis. This approach is significant in comparison with the great quantity and the diversity of the data, as well as the need of a personalized treatment for each patient.

The constraints of this work concern in particular to the need for an approach focused on the characterization and determination of the parameters of specific ECG signal of each person. The habits of daily life as well as physiological characteristics vary according to individuals. The complexity of the problem resides in a great inter-individual variability of the recorded data, and also in broad intra-individual possible modifications being given the often not very foreseeable aspect of human behaviors [2, 3].

The problem also is posed on the level implementation of an efficient algorithm intended to solve all the preceding constraints and adapted to mobile phones. This implementation requires many constraints (low resource calculation, memory capacity, resolution…) to run properly.

In this context, the considered study leads to an inexpensive solution, efficient and comfortable for patients at any time and anywhere, provided that they have a mobile terminal. Indeed, they could benefit of a medical monitoring security, without the inconvenience and without excessive expenditures.

8.3   Our Vision

The cost of the health represents a considerable weight in the economic balance sheet on international scale. Also, in many countries, the aging or psychological shocks of the population tends to increase the number of people strongly requiring medical monitoring even more or less intensive care and of this fact the global cost of medical care.

As all the technologies, the mobile telephony is evolved, and actually the offered possibilities are more important than ten years ago. But indeed as often, the majority of users use only the basic functions, phoning and sending messages (SMS), what have already allowed envisaging multitude applications.

In our research task on the subject, we propose the exploitation of the mobile phone in other applications apart from the vocal communications.

The idea is to divert these devices of their basic function to make them useful for medical supervision.

Whereas ten years before, such a taking off would have required large means as well as a large infrastructure, today a simple mobile phone can effectively contribute to the protection of human lives.

8.3.1   Platform System and Functions

The considered system allows a patient to be in contact, at any time, with his doctor for medical monitoring (Figure 8.1).

The implemented application on a mobile phone, functions on all mobile terminals or PDAs equipped with a J2ME virtual machine. This algorithm allows communications GSM / GPRS with other devices. In this case, the transmitted data concerns the ECG signal collected on remote sites. The procedure consists to implement on a mobile device the following operations:

— Display of 5000 samples (amplitudes: mV) of signal ECG of the patient.

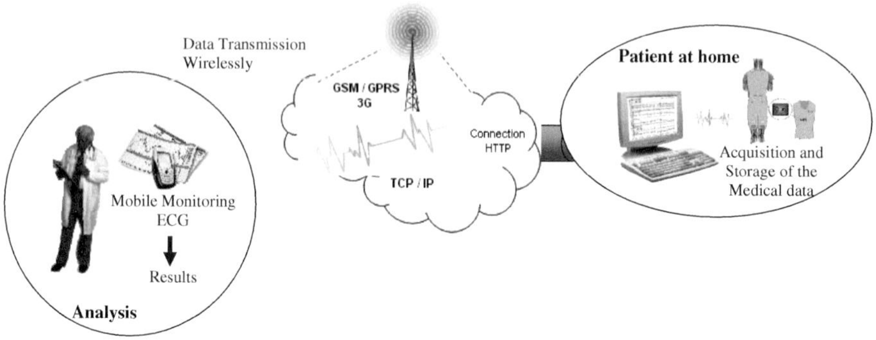

Fig. 8.1 Architecture of the platform system.

— Detection of the peaks R (times between each two peaks, cardiac frequency).
— Calculation of QRS durations.
— Visualization of ECG signal with 5000 samples on the screen of mobile terminal.
— Possibility to zoom the ECG signal.

The first part which must be realized concerns the collection of remotely medical data on the mobile terminal. These data are generated by sensors installed at patient's home.

Then it is necessary to make connection between the medical supervision device (mobile phone) and the remote monitoring system at patient's home, so that they can exchange their data.

An adaptation strategy to medical context was followed to generate analysis results of calculation algorithm implemented on doctor's mobile terminal (detail in the following sections). Thus, the doctor is able to check the result of the diagnosis obtained by this algorithm.

8.3.2 Choice of Technology

The analysis carried out made it possible to better understand the principal protocols which intervene in the development of our application.

After studying the various technologies, in terms of remote exploitation of the stored data on a computer at home, the most adapted solution consists to use a simple mobile phone linked by a GSM / GPRS to this database. This technology is easy and rapid to be implemented.

This implies the existence of an http connection between doctor's mobile terminal and monitoring home automation system via a WAP gateway to recover data of ECG signal. This choice is dictated by the following characteristics: [4]

— Http is obligatorily implemented on all terminals MIDP (J2ME).
— Http is independent of the network.
— The port 80 of the http protocol is more easily working on the firewall.
— Http protocol is implemented by default in J2ME package. Other protocols are not necessarily available.

The transfer of the medical data of a patient at home on the mobile phone is based on a communication WSP/ Http. As its name suggests Wireless Session Protocol (WSP), session layer allows the connection setting to make transactions. Thus it allows the layer application to profit from two different types of sessions:

— Connected session mod which the layer session will interact with the layer transaction.
— No-Connected session mod where the session layer will act directly at the transport layer for sending brutes' datagram.

WSP as a whole is equivalent to the http protocol. We find moreover many identical implementations to the http in WSP.

8.3.3   Choice of Development Environments

The recording and the acquisition of the ECG data were carried out using MATLAB.

Java applications have been implemented under NetBeans IDE environment.

A simulation tool Sun Java ™ Wireless Toolkit 2.5 for CLDC was exploited to examine all the possible wireless communications. It allows applications on devices with low calculation resources such as a mobile phone [5].

The choice of Java is justified by the different problems associated to coding in C ++ on Symbian operating system:

— Management of the memory: for the majority of applications, Java system seems to be sufficient.
— Environment of execution: the proposed options on executable Java as protections for downloading or secure execution are free, whereas C ++ it is necessary to develop them, test them and maintain them.
— Perpetuity: Java seems to have been accepted for the development of applications on mobile phones. The future developments will make Java perhaps as fast as C ++.

Thus for that principal reason, Java was chosen in our project, but it is necessary to mention that both environments can be used [6].

8.4 Implementation of Proposed Model

As mentioned before, our implementation achieves the medical service which provides the continuity of remote monitoring at home and immediate alarms to deal with the patient in the event of need.

The schedule of conditions of our project consists of:

— The implementation of this service requires the development of two distinct applications:
  • A first to be installed on the mobile phone to detect and treat the critical situations of patient via wireless support.
  • A second function on the PC at home in order to acquire and record data relating to the patient.
— To program the application in a language which is most portable possible, the algorithm must be simple to use and install.
— To program a user interface of high quality.

The suggested model is also based on techniques of programming adapted not only to the limited resources (devices of medical supervision) but also to the generated heterogeneous parameters. What allows in particular showing the diversity of profiles of persons and types of generated situations, including the simulation of "normal" modifications and disturbing of behavior.

The following paragraphs present (1) Development on mobile phone, (2) Development on PC, (3) the simulation of the proposed model.

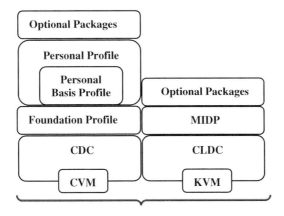

Fig. 8.2 Architecture of J2ME.

8.4.1 Development on Mobile Phone

J2ME is a collection of technologies and specifications which are conceived for various parts of the market of the small devices. The principal part of platform J2ME is composed of two different configurations (Figure 8.2): Connected Device Configuration (CDC) and Connected Limited Device Configuration (CLDC) [4].

A configuration defines the central libraries of Java technology and virtual storage capacities of the device. CLDC is adapted to recent mobile phones. This configuration is useful for our application. To still note, that in the case of J2ME, the virtual machine is called KVM for Kilobyte Virtual Machine.

At the top of the configurations (Figure 8.2), there are the profiles which define the functionalities in each specific category of devices. The "Mobile Information Device Profile" (MIDP) is a profile for the mobile devices using configuration CLDC, like the mobile phones. Profile MIDP specifies the functionalities like the use of the interface user, the persistence of storage, the setting in network and the model of application.

On the majority of the current phones, J2ME is composed of configuration CLDC and profile MIDP.

In addition to standard MIDP (Figure 8.2), additional (optional) packages can be added according to the devices, allowing the use of their specificities.

As these options are typically reserved to mobile phones, it was natural to not integrate them directly in the profile MIDP.

So, the development of our application on mobile phones is based on the use of configuration CLDC and profile MIDP. In addition to these two standard elements, we have exploited some optional packages such as WMA for the management of services SMS / MMS and Web Services API. The libraries necessary for the implementation for each component of J2ME are as follows: [6]

API MIDP: Is, till now, what is installed on mobiles "compatible J2ME":

-javax.microedition.lcdui: It provides the graphic components necessary to the creation of applications.

-javax.microedition.midlet: It provides the component application as well as the primitives managing the life of the application.

-javax.microedition.rms : A possibility of storage of information on the terminal.

API CLDC:

Javax.microedition.io: It contains the classes making it possible to connect itself via *TCP/IP* or *UDP*. The principal object of this package is the class *Connector* [4].

This network part determines which means used to communicate medical data.

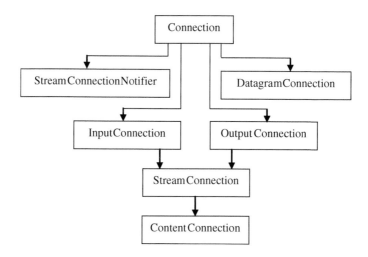

Fig. 8.3 Arborescence of the classes javax.microedition.io.

The above diagram corresponds to the different classes javax. microedition.io. It is thanks to these implementations that we can make http requests.

The application on the client, called MIDlet, is carried out with the virtual machine J2ME (KVM) on the mobile terminal. It has the role to receive measurements of the devices of the patient at home, to treat these data or to store them if necessary.

It also allows to the doctor to send an alarm in case of a critical situation.

8.4.2 Development on PC

The second application (patient at home) is composed of different subprograms. They have for function to store the entering data in a database, and allow to the doctor to establish the diagnosis of the patient either by execution of an algorithm incorporated into his cell phone, or by exploiting the received data.

8.4.3 Simulation of Proposed Model

The networks GSM/GPRS are useful to transmit ECG data of a patient. Currently, mobile phones of last generations are able to send and receive all sorts of messages (text, image, sound…). They offer in addition to the voice communication, a supply of services on a large scale, which allows considering a multitude of applications for these devices.

Our investigation consists to integrate an ECG signal on a mobile phone to ensure a medical remote monitoring service of patients at home. We describe in what follows the basic concept in electrophysiology of the heart in order to specify the components of the electrocardiogram (ECG).

8.4.3.1 Electrocardiography

Electrocardiography deals with the study of the electrical activity of the heart muscles.

The human body being electrically conductive, the electric potentials produced by the activity of the heart can be collected by electrodes placed on the thorax the model recorded of this electric activity of the heart, on a frontal

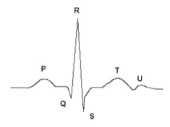

Fig. 8.4 The complete cardiac cycle.

Table 8.1 The parameters of an ECG signal.

Type of wave	Origin	Amplitu-de (mV)	Durati-on (Sec)
Wave P	Atrial depolarization or contraction	<= 0.2mV	Interval: P-R 0. 12 À 0.22
Wave R	Repolarisation of the atria and the depolarization of the ventricles	1.60	0.07 À 0.1
Wave T	Ventricular repolarisation (Relaxation of myocardium)	0.1 – 0.5	Interval: Q-T 0.35 À 0.44
Inter-val S-T	Ventricular contraction		Interval: S-T 0.015 À 0.5
Wave U	Slow repolarisation of the interventricular (Purkinje fibers) system.	< 0.1	Interval: T-U 0.2

level (derivations of the members) and on a horizontal level (derivations of precordiales) is an electrocardiogram.

For each heartbeat, the electrocardiogram record 4 successive waves (P, R, T, U) (Figure 8.4).

Each of these waves is characterized by its amplitude and its duration. The Table 8.1 summarizes these values for a normal person [7].

This diagnostic tool allows detecting the rhythmic cardiac pathologies, muscular, extracardiaques metabolic problems, Medicinal, hémodynamiques and others.

8.4.3.2   Application

This section describes a medical application implemented in a mobile phone to treat and characterize the ECG signal.

Our application consists to develop a MIDlet to take remotely the evolution of the state of patients and to calculate parameters witch characterize the ECG signal.

The implementation of the proposed model for transmission simulating, storage and data processing of the electrocardiogram is realized with the J2ME environment. For reasons relating to the rapid evolution of technology, it is always preferable to avoid carrying out specific applications to a type of mobile equipment owner (Windows, Symbian, Palm / OS) [8]. J2ME allows the development of applications which can run on all compatible mobiles.

The following paragraphs present the principle of the implementation, the global structure of the implementation of simulation and finally the analysis result of the calculated parameters (Cardiac frequency averages, QRS duration...).

8.4.3.2.1 *The principle of implementation*

Being given the complexity between the portable telephone's technology and number of parameters of the model which must be defined in priori, an adaptation of data to this context consists to use a set of files in the format text for the definition of the current exploited values for the simulation and the default values. A graphical interface on the phone allows the display of contained parameters in the files. The simulation, once launched, can then retrieve values from these files. At the end of the application, the generated results are stored in memory of the mobile terminal, and possibly displayed on the screen.

Figure 8.5 presents the general principle of the process implementation of simulation.

The result interface is presented on Figure 8.8 and Figure 8.9. The parameters are calculated according to the simulation model sequences on which they involve.

8.4.3.2.2 *Global structure of the implementation of simulation*

The global structure of the simulation program is completely sequential. It calls, one after the other, the functions realizing the principle stages of the simulation and successively corresponding to the generation of the different parameters of ECG signal.

These parameters include the duration between two successive peaks R, complex duration QRS and cardiac frequency. Each called function takes as

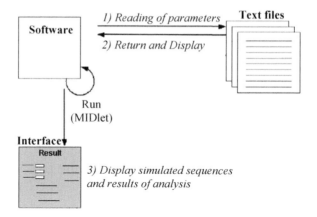

Fig. 8.5 Principle of the process implementation of simulation.

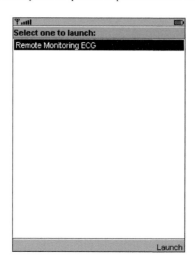

Fig. 8.6 Launch of the algorithm.

entering parameter the results of the call of the previous function and provides the results of its execution to the following function.

The stages of the MIDlet execution are detailed in the following section.

8.4.3.2.3 *Calculated parameters and analytical result*

As we have already seen before, J2ME wireless development was exploited to the implementation of the proposed simulation model of ECG signal on a mobile phone.

In this section, we present the various stages of execution of the algorithm. All this series of tests was made thanks to the phone emulator.

During the launching of the application, MIDlet allows to the doctor to activate the mode of the medical remote monitoring ECG (Remote Monitoring ECG):

At the beginning, the application will operate and communicate in autonomous mode with measurement devices. Then, the phone collects 5000 samples, generated by the electrodes placed on body of the patient at home (Figures 8.7 and 8.8) and stores these data.

Fig. 8.7 Acquisition of 5000 samples of a patient (ECG signal).

ECG Data:	
Sample Number	Amplitude
1	1184
2	1181
3	1192
4	1203
5	1223
6	1248
7	1240
8	1253
9	1235
10	1222
11	1210
12	1192
13	1171
Exit	Result

Fig. 8.8 Data ECG transmitted to the mobile phones.

These samples are points in mV which constitute the Electrocardiogram.

Figure 8.8 shows the organization in vectors of the ECG values transferred via GSM/GPRS to the internal memory of the mobile phone. They are obtained after amplification at domicile of patients with a gain factor of 1000 and a sampling frequency of 128 KHz.

Such a medical application proposes a set of services to the health's professionals (list of the patients, the display of the medical profile of a patient…). These services make treatments with variable complexities (management of data via a data base, numeral calculation…) and exchange data with the user through a graphical interface on a mobile terminal. This environment type presents important and heterogeneous information, a great variability and numerous possibility of evolution. Indeed, the offered resources on the level of terminal can be extremely different according to the use of a personal assistant, a laptop or a workstation. Thus it is necessary to implement an adaptation strategy to conceive and develop the algorithm by respecting these prerequired constraints.

The graphical interface which is in this case the portable's screen, allows to collect, display, store and calculate the parameters of simulation (time between two successive peaks R, QRS duration,…), after an adaptation of medical data to the context. Our adaptation strategy consists to convert or replace the content of a text file to be sent to the doctor.

For example, we convert a character to a number because the values of the ECG are recorded as format ".Txt".

The principle of this process is the characterization of ECG signal. After receiving the medical data (Figure 8.8), the doctor is invited to consult the latest results of his patient (Figure 8.9) in order to take the adapted decision.

The diagnosis and treatment can be done using the implementation of calculation algorithm on the mobile. It allows calculating the most significant parameters necessary to the characterization of an ECG.

First of all the incorporated algorithm in the mobile phone must position all the peaks R (after an adaptation of recorded data in the textual files quoted before). It calculates the duration between each two successive peaks R by multiplying the number of point located between these two peaks by the sampling frequency. To deduce the average cardiac frequency it is enough that it calculates the average of the durations of a fixed number of peaks R than

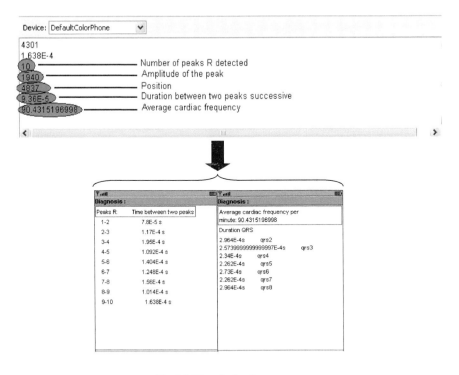

Fig. 8.9 The calculated parameters.

it divides one fixed period (one minute in this case) by this average value (Figure 8.9).

The calculation of the QRS duration follows the same procedure. The last stage is the display of ECG signal on the mobile terminal's screen (Figure 8.10).

In this scenario, it has involved several peaks R for the implementation of each model containing the definition of parameters. Each simulation model containing ECG data corresponding to a patient generates parameters which characterize the patient: time between two successive peaks, duration QRS and cardiac frequency. Thus, the doctor can observe the ECG signal in real time on his screen with 5000 samples: (Figure 8.10).

A more importantly option allowing the zoom of the part which presents an anomaly, is implemented in our application. It is enough to the doctor to introduce a begin point and an end point (time interval) in order to widen the part in question (Figure 8.11).

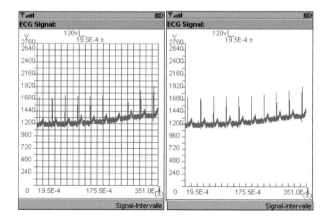

Fig. 8.10 Show ECG waveform of the patient (with/without grid).

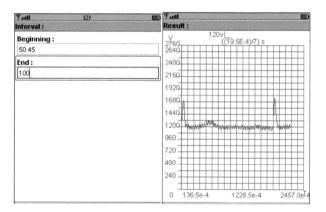

Fig. 8.11 Zoom part of the curve.

At present we are interesting in the transmission of an ECG signal on several channels to make a multitude of analysis for the same cardiac patient and increase so the efficiency of the system, while having the possibility of revising all the results.

8.4.3.2.4 *Comparison with the simulation MATLAB*

The sequence data generated for the parameters of ECG in the context of medical remote monitoring (MIDlet) are validated by a comparison with simulated data under MATLAB environment (Figure 8.12).

The validation which is carried out about the implemented simulation process by the tool of J2ME development makes this algorithm to be exact in

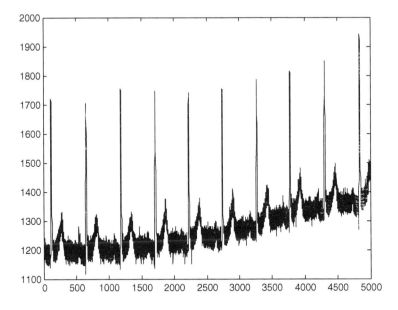

Fig. 8.12 ECG signal.

terms of calculation, powerful and effective. It was developed in the context of medical remote monitoring, from the recorded data sequences at home. The obtained results allow making advance the strengths points and the innovative qualities of this algorithm compared to other carried out works in similar domains or related, at the same time our algorithm presents better perspectives of improvement of its efficiency.

8.4.4 Quality of Service

Within the framework of this work, the problem of the quality of service of the mobile networks for the second or third generation, lives on one hand in a environment with various requirements of good functioning and on the other hand in two ways of transport of the information: circuit or packet.

The doctor wants to know always about state of evolution of the patient. What requires to have at least an acceptable quality of service in the mobile communication to be able to sanction by a correct decision-making.

We can say that the zero risk does not exist. Prevention measures in the system of remote monitoring, chosen suitably according to needs, can reduce considerably this risk. For that purpose, this application requires a real-time transmission and is not tolerant in the errors. It requests for more or less raised debits.

8.4.5 Instructions of the Algorithm

8.4.5.1 Minimal configuration

To be able to use the algorithm of analysis, one needs a mobile phone, which has the following characteristics:

— Support of profiles MIDP 2.0 or more and of CLDC 1.0 (or CLDC 1.1).
— Support of the api WMA 1.1 (JSR 120) and Web Services API (JSR 172).
— 60ko minimum of memory capacity available on the phone to store 5000 samples and the calculated parameters (after analysis).

8.4.5.2 Diffusion of software

To diffuse this software, there are several possibilities, but in each case, the two only files which must be provided to the user are: the file ".jad" and the file ".jar". These files must be transferred on the mobile phone, for that there are several solutions:

— Download on the phone of an E-mail containing both files in attached peaces.
— Access to the files placed on a Web server.
— Transfer using Bluetooth or infrared signal.
— …

In all the cases, the two files must be in the same one repertory or the same E-mail.

8.4.5.3 Installation of the program

Being given that the installation differs according to the mobile phones, it is not possible to give a precise procedure. But normally, the simple fact of opening the file ".jad" is enough to install the application.

8.5 Evaluation

The proposed simulation algorithm is articulated on two fundamental points: the first one relates to the simulation step in the respect of the complexity and objectives of the medical remote monitoring context at home, the second one is the more global vision in the resolution cycle of construction problem of the behavior profile of person to ensure a critical situation. The suggested remote monitoring consists to monitor and diagnose the state of a patient using the methodology developed in this project. Thus, the doctor treating a person with risk cardiac can at any time control the state of his patient by consulting in real-time the ECG on his mobile terminal and the classification of pathology by the developed algorithm.

8.6 Conclusion

This article refers to analysis biological signals of a patient, recorded at home and detected on the mobile phone of his doctor. This technique allows medical remote monitoring of cardiac or hypertensive patients.

Also, the identification, by the developed algorithm, of the medical profile of a patient at home and the detection of critical situations cannot cover all medical indicators corresponding to each patient.

Thus, the improvements of this algorithm must be adapted to diagnose new pathologies.

This solution, not costly and easily realizable, is adapted to the portable devices ensuring medical monitoring anytime and anywhere. It is in this vision that other services, associated to mobiles and intended for the telemedicine and the house automation are under development.

Acknowledgement

I deeply thank my supervisor and the person in charge of the laboratory STIC, Mohamed FEHAM, Professor at the university Abou Bekr Belkaid of Tlemcen, Algeria. The correctness of his advice, the motivation and the project financing. They were very precious and brought a successful conclusion to this work. A special thank to the researchers of the laboratory who gave their help to the realization of this project.

References

[1] O. Fouial, I. Demeure, "Fourniture de services adaptables dans les environnements mobiles". Proceedings of European Conference of Systems with adaptable and extensible components. Grenoble, France, 2004.

[2] B.G. Celler, T. Hesketh, W. Earnshaw, E. Ilsar, "An instrumentation system for the remote monitoring of changes in functional health status of the elderly at home ", in Proc. of the 16th Annual IEEE Engineering in Medicine and Biology Society, Baltimore, USA, 1994, pp. 908–909.

[3] L. Chwif, R.J. Paul, "On simulation model complexity", in Proc. of the 32nd conference on Winter simulation, Orlando, Florida, 2000, pp. 449–455.

[4] Delb. B, J2ME, Application java pour terminaux mobiles, EYROLLES, Paris, France, 2002.

[5] Quintas. A. F, "Bluetooth J2ME Java 2 micro edition", manual de usuario y tutorial, Madrid: Ra-Ma, 2004.

[6] Knudsen. J, "Wireless Java Developing with J2ME", Second Edition, Apress, Berkeley, United States, 2003.

[7] R Legameta, P.S. Addisson, N. Grubb, CE. Robertson, K. Fox, J.N. Watson, "Real-Time Classification of ECGs on a PDA", IEEE Trans. On Information Technology in Biomedicine, vol. 30, 2003, pp. 565–568.

[8] Mahmoud. H, "Learning Wireless Java", O'Reilly, Sebastopol, USA, 2002.

9

Statistical Validation of Physiological Indicators for Non-invasive and Hybrid Driver Drowsiness Detection System

Eugene Zilberg*, Zheng Ming Xu*, David Burton*,
Murad Karrar* and Saroj Lal[†]

*Compumedics Medical Innovation Pty Ltd, Australia
[†] University of Technology Sydney, Australia

9.1 Introduction

Driver drowsiness is recognised as an important factor in the motor vehicle accidents [4, 10, 15]. It has been shown that a number of physiological indicators such as the spectral content of electroencephalogram (EEG) [9, 11, 12, 17], blink rate [20], individual blink parameters [2] and degree of eye closure (PERCLOS) [5] are associated with the increase in drowsiness [10] and can respectively be used for detecting driver drowsiness.

Compumedics Medical Innovation Pty Ltd, Australia proposed application of non-invasive piezofilm movement sensors that can be incorporated into car seat, seat belt and steering wheel [1]. These sensors are potentially capable of recording patterns of driver's movements, breathing and even heart rate that could be used for identifying the level of drowsiness. Another aspect of Compumedics patented technology includes integration of different kinds of signal analysis including morphological processing of EEG and eye movement patterns that was successfully used for automatic analysis of sleep recordings [16].

The objective of this article is to establish statistical associations between increase in the driver drowsiness and the non-invasive physiological indicators proposed by Compumedics as well as to demonstrate increase in the strength of these associations in the hybrid system when a "gold standard" indicator (EEG spectral content) is combined with the non-invasive piezofilm movement sensors embedded in the car seat. The identified statistical associations can serve as a foundation for designing the vehicle-based fatigue countermeasure device as well as highlight potential difficulties and limitations of detection algorithm for such devices.

This article follows the earlier investigations [24, 25] that described the driver simulator study conducted as a part of the ARC Linkage grant with University of Technology Sydney, Australia, to explore possibilities offered by the Compumedics non-invasive and hybrid technologies, introduced the observer driver drowsiness scale, explained the data analysis methodology and presented results of statistical analysis focusing on investigation of associations between transition to a state of drowsiness and patterns of selected physiological indicators including the surface movements of the seat and steering wheel, EEG and durations of eye movements.

9.2 Methods

9.2.1 Car Simulator Study and Data Selection

A car simulator study comprising 60 non-professional fully licensed drivers was conducted at the Monash University Accident Research Centre (MUARC) in 2005–2006 [24, 25]. The objective of this study was to record multitude of physiological signals that could be potentially used as components in the driver drowsiness detection algorithm as well as a number of measures that could be employed as independent indicators of the drowsiness level. The recorded physiological signals included 10 EEG channels, chin EMG (electromyogram), ECG (electrocardiogram), left and right EOG (electrooculogram), eyelid movement sensor, thoracic respiratory band, 8 steering wheel pressure gauge signals and multiple signals from the piezofilm sensors located on the car seat and steering wheel. The movement sensors included 7 sensors on the steering wheel, 5 on the back of the seat and 5 on the bottom section of the seat. Siesta portable sleep diagnostic recording system (Compumedics Pty Ltd, Australia) was used for physiological data recording. The information for the independent assessment of the drowsiness level included 4 video signals, estimate of PERCLOS [5] from the FaceLAB system [7] as well as a list of driving performance parameters. The participants also filled a number of pre- and post-study questionnaires such as Lifestyle Questionnaire [4], profile of mood states [14], and anxiety state-trait [19].

The developed observer drowsiness rating scale [24] included 5 levels similar to those described in [22] – from alert to extremely drowsy and was based on observing a number of indicators including duration and rate of eye blinks and other eye lid movements, degree and duration of eye closures, direction and focus of eye gaze, patterns of facial, hand and other body movements, yawning etc. The appropriate dangerous level of drowsiness was selected as "significantly drowsy" (level 3) based on appearance of eye closure episodes.

9.2.2 Statistical Associations in Question

In [24, 25] we conducted statistical analyses of the time courses of different physiological indicators during episodes of transitions to the state of significant drowsiness. The main findings of these analyses were

- There was a pattern of reduction in the magnitude of seat movements associated with transition to the state of significant drowsiness with the sensors on the back of the seat demonstrating more significant association than those on the bottom of the seat;
- There was a pattern of increase in the percentage of EEG alpha activity associated with transition to the state of significant drowsiness for the central and occipital EEG derivations with the former derivation demonstrating more significant association when correlation is taken into account;
- There was a pattern of increase in the duration of eye movements associated with transition to the state of significant drowsiness with eye movements scored from the frontal EEG. The eye movement duration calculated from the eyelid movement sensor signal demonstrating more significant association than that scored from EEG (although this finding is based on a very small subset of participants);

- There was a pattern of reduction in the magnitude of steering wheel surface movement and pressure associated with transition to the state of significant drowsiness with the movement sensors demonstrating significant association in contrast to the pressure and both measures having substantially less significant associations than the seat movement.

Although the method of analysis of the time course of physiological indicators during transition to drowsiness state by means of fitting linear regression models with the correlated observations [3, 18, 21] enabled to establish the described associations there were still a number of important questions that it could not help to answer

- What is the association between physiological indicators and probability of a predetermined state of drowsiness?
- How can different physiological indicators be compared in relation to their association with the probability of drowsiness?
- How can different physiological indicators be combined to increase the strength of association with the probability of drowsiness and what is the contribution of individual measures into the increased association?

Fitting the binary (and potentially nominal/ordinal logistic regression models [6]) can potentially address these issues.

9.2.3 Statistical Models

The most general approach to define the outcome variable would be to use the same set of categories as in the visual scale of drowsiness described in [24, 25]. However as a starting point for analysis covered in this article we employed a binary model with the state of drowsiness being defined as the drowsiness level of 3 (significant drowsiness) or higher while the state of alertness being defined as any drowsiness level below 3.

Only the observations recorded during the episodes of transition to drowsiness were used in this analysis. The rationale for this is that initially we had to establish associations that are specific for those most important episodes unaffected by other factors that could be present during states of prolonged alertness or deeper drowsiness. If no statistically significant associations are found for the episodes of transition to drowsiness the expansion of this method to longer time intervals would be pointless.

The observation values for covariates for this analysis are formed in the following way. All episodes of transition to significant drowsiness are divided into non-overlapping 20 second segments. The selection of 20 s interval was a trade-off between 30 s intervals employed in [24] to analyse the seat movement and eye movement duration and 10 s intervals used for he EEG analysis. For every segment the following values are calculated

- The maximum drowsiness rating out of the two 10 second intervals that comprise the 20 second segement;
- Seat piezofilm movement sensor data for all 10 seat sensors calculated as explained in the section on the time course analysis but averaged over 20 s segments;
- C4 and O2 EEG derivation data with the maximum 2 s alpha band percentage calculated over 20 s segments;
- Eye movement duration data derived from the frontal EEG and averaged over 20 s segments.

The total number of observations with valid non-zero values of all covariates for 115 episodes of transition to drowsiness was 1029.

If correlation between observations is ignored the binary logistic model for the log odds of drowsiness and a combination of N covariates can be formulated in a general form [20] as

$$\log \left[\frac{\Pr\{Y_i\}}{1 - \Pr\{Y_i\}} \right] = \beta_0 + \sum_{k=1}^{N} \beta_k COVAR_{ki} \tag{9.1}$$

where

- Y_i - binary state of drowsiness for the ith observation;
- $COVAR_{ki}$ - value of the kth covariate for the ith observation;
- β_0, β_k - regression coefficients to be estimated with the coefficient β_k representing the effect of the kth covariate.

The graphical example of fitting the model (9.1) for the case of central EEG derivation C4 as a single covariate is presented in Figure 9.1. To generate

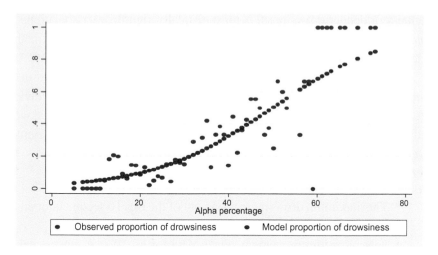

Fig. 9.1 Fitting binary logistic regression model for association between alpha band percentage for the central EEG derivation and odds of drowsiness.

this graph all 1029 observations were put into one of 100 bins based on the respective nearest integer to the value of the C4 alpha percentage. For every bin the observed proportion of observations with significant drowsiness was calculated. Then the univariate binary logistic regression model was fitted and the predicted proportions of drowsiness were generated for every bin. It is evident that binary logistic model is an adequate method for description of association between increase of the alpha percentage and odds of drowsiness.

The set of covariates in (9.1) was expanded to include the observations of respective physiological parameters at the preceding 20 s segments (that will be denoted with index *previous*) as well observations referenced to the value in the start of a respective transition episode (trajectory) segments (that will be denoted with the additional index *ref*). The observations of covariates at the given 20 s segment will be denoted by the index *recent*. Finally the observations for the preceding 20 s second segment referenced to the start of the trajectory will be denoted with the index *previous_ref*. The preceding segments are included in the regression model to investigate the predictive capabilities of different physiological indicators while referencing to the start of a trajectory is important for the covariates that have no physiological interpretation for their absolute values (like seat movement magnitude). This notation can be demonstrated with the following formulae for an arbitrary covariate X at the time denoted by the index j and the initial 20 s interval of a transition episode having index 0.

$$X\_recent = X_j \tag{9.2}$$

$$X\_recent\_ref = X_j - X_0 \tag{9.3}$$

$$X\_pevious = X_{j-1} \tag{9.4}$$

$$X\_previous\_ref = X_{j-1} - X_0 \tag{9.5}$$

Correlation between observations can be taken into account by means of fitting the random-effects non-linear mixed model [8, 23]

$$\log\left[\frac{\Pr\{Y_{ij}\}}{1 - \Pr\{Y_{ij}\}}\right] = \beta_0 + b_i + \sum_{k=1}^{N} \beta_k COVAR_{kij} \tag{9.6}$$

where

- Y_{ij} - binary state of drowsiness for the jth observation of the ith cluster (eg. trajectory);
- $COVAR_{kij}$ - value of the kth covariate for the jth observation of the ith cluster;
- b_i - zero-mean normal error component for the ith cluster.

For example the model (9.6) that has the most EEG C4 alpha percentage for the 20 s interval identical to the one where the binary state of drowsiness is estimated, the eye movement duration from the previous 20 s interval and the references signal from the first seat for the current 20 s episode can be expressed in the proposed notation as

$$\log\left[\frac{\Pr\{Y_{ij}\}}{1 - \Pr\{Y_{ij}\}}\right] = \beta_0 + b_i + \beta_1 \times C4\_recent + \beta_2 \times Eye\_previous$$

$$+\beta_1 \times Seat1\_recent\_ref$$

$$= \beta_0 + b_i + \beta_1 \times C4_{ij} + \beta_2 \times Eye_{i,j-1}$$

$$+\beta_1 \times \left(Seat1_{ij} - Seat1_{i0}\right) \tag{9.7}$$

The models were ranked by means of comparing their log-likelihood statistics and deviances and the objective was to obtain the parsimonious (simplest statistically significant) models for the given combinations of physiological indicators. The diagnostic capabilities of the fitted combinations of covariates were indicatively compared by means of estimating the areas under the respective ROC curves.

The analysis started with the EEG alpha percentage that is known to be an important contributor into EEG based drowsiness rating [9, 11, 12, 17]. Then diagnostic capabilities of other variables and combinations of variables were investigated with performance of the EEG based measure used as a benchmark.

9.3   Results

9.3.1   EEG

Table 9.1 presents the best univariate models (1, 6) for the central and occipital EEG derivations and different correlation assumptions. It was also found that the most recent EEG observations provide better fit that the observations at the previous 20 s segment therefore EEG does not have the predictive capability. For the occipital derivation referencing against the value at the start of transitions episode achieved a better model while this was not the case

Table 9.1 Univariate binary logistic regression for central and occipital EEG derivations as covariates and log odds of drowsiness outcome with different correlation assumptions.

Correlation model	Cental EEG derivation (C4-A1) *recent*	Occipital EEG derivation (O2-A1) *recent_ref*
Independent observations		
Regression coeff.	7.45 [6.00; 8.90]	7.16 [5.79; 8.53]
z	10.08	10.23
p-value	< 0.001	< 0.001
Log-likelihood	−438.03	−434.05
Within same transition episode (random effects)		
Regression coeff.	9.49 [7.50; 11.48]	8.74 [6.89; 10.60]
z	9.34	9.22
p-value	< 0.001	< 0.001
Log-likelihood	−424.43	−424.35
Between all observations for subject (random effects)		
Regression coeff.	10.98[8.87; 13.09]	7.59 [6.04; 9.15]
z	10.21	9.56
p-value	< 0.001	< 0.001
Log-likelihood	−414.85	−431.30

Key: *p*-value= significance level, z is the ratio of estimate to its standard deviation.

for the central derivation. It is also evident that for the assumption of independent observations the occipital EEG provides the better model while once a greater degree of correlation taken into account the model for the central EEG derivation starts to fit better. There is increase in the estimated absolute values and variances of regression coefficients when correlation is taken into account. This could be attributable to the difference in interpretations of models (9.1) and (9.6) that refer to "population-average" and "subject-specific" effects [8]

The parsimonious models with multiple EEG parameters are presented in Table 9.2. For different correlation assumption different combination of four EEG parameters were found to maximise the log-likelihood. In all cases the estimated values of all regression coefficients were positive as expected. It appears that combination of central and occipital EEG derivations as well as

Table 9.2 Parsimonious binary logistic regression models for central and occipital EEG derivations as covariates and log odds of drowsiness outcome with different correlation assumptions.

Correlation model	Covariate 1	Covariate 2	Covariate 3	Covariate 4
Independent observations				
Covariate	C4_recent	C4_previous_ref	O2_recent_ref	O2_previous
Regression coeff.	3.67 [1.87; 5.47]	1.84 [0.27; 3.38]	3.77 [2.09; 5.45]	3.57 [2.07; 5.07]
z	4.00	2.34	4.40	4.67
p-value	< 0.001	0.019	< 0.001	< 0.001
Log-likelihood		−399.72		
A-ROC		0.765 [0.726; 0.804]		
Within same transition episode (random effects)				
Covariate	C4_recent_ref	C4_previous	O2_recent_ref	O2_previous
Regression coeff.	5.26 [2.72; 7.79]	3.96 [1.39; 6.53]	5.12 [2.69; 7.55]	5.14 [2.68; 7.60]
z	4.06	3.01	4.13	4.09
p-value	< 0.001	0.003	< 0.001	< 0.001
Log-likelihood		−382.74		
A-ROC		0.770 [0.0.731; 0.808]		
Between all observations for subject (random effects)				
Covariate	C4_recent	C4_previous	O2_recent_ref	O2_previous
Regression coeff.	7.14 [4.67; 9.61]	4.29 [1.72; 6.86]	4.64 [2.71; 6.56]	4.98 [2.88; 7.08]
z	5.67	3.27	4.71	4.64
p-value	< 0.001	0.001	< 0.001	< 0.001
Log-likelihood		−370.34		
A-ROC		0.756 [0.715; 0.796]		

Key: C4- central site, O2- occipital site, A-ROC is the area under Relative Operating Characteristic (ROC) curve, p-value = significance level, z is the ratio of estimate to its standard deviation.

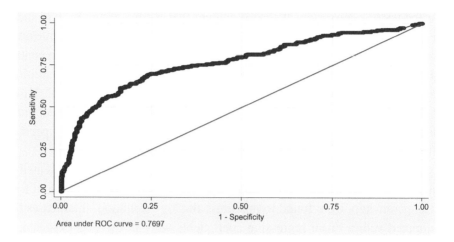

Fig. 9.2 ROC curve for the parsimonious model with the central and occipital EEG sensor signals as parameters.

adding the observation for preceding time interval improved fit of the respective regression models.

The final log-likelihoods of -399.42, -382.74 and -370.74 will be used as respective benchmark values in the course of the further analyses involving other physiological variables. Finally the typical ROC for the combination of EEG covariates estimated for the random-effects model (9.2) and within-transition correlation is displayed in Figure 9.2.

9.3.2 Piezofilm Seat Movement Sensors

Table 9.3 presents results of fitting univariate models (1, 6) for the first seat sensor only. It is evident that the preceding observation provides better fit for all correlation assumptions. This was also the case for other seat sensors therefore it is evident that the seat movement data have predictive capability unlike the EEG derived covariates. This conclusion appears plausible from the physiological prospective as burst of alpha activity are likely to manifest more advanced stage of drowsiness than reduction in movement magnitude. The increases in absolute values and variances of regression coefficients when correlation is taken into account again could be attributable to the difference in interpretation of models (9.1) and (9.6). The log-likelihoods for individual

Table 9.3 Univariate binary logistic regression for a selected seat movement sensor as a covariate and log odds of drowsiness outcome with different correlation assumptions.

Correlation model	peak to peak deflection over 1 s averaged within latest 20 s interval	peak to peak deflection over 1 s averaged within preceding 20 s interval
Independent observations		
Regression coeff.	−59.397 [−79.070; −39.724]	−67.807 [−89.091; −46.523]
z	−5.92	−6.24
p-value	< 0.001	< 0.001
Log-likelihood	−475.195	−473.225
Within same transition episode (random effects)		
Regression coeff.	−66.232 [−90.640; −41.824]	−75.041 [−101.144; −48.939]
z	−5.32	−5.63
p-value	< 0.001	< 0.001
Log-likelihood	−470.576	−468.294
Between all observations for subject (random effects)		
Regression coeff.	−61.027 [−81.681; −40.372]	−70.455 [−92.901; −48.008]
z	−5.79	−6.15
p-value	< 0.001	< 0.001
Log-likelihood	−471.019	−468.125

Key: p-value= significance level, z is the ratio of estimate to its standard deviation.

seat movement sensors are well below the respective values for the univariate EEG models presented in Table 9.1.

The results of fitting models (6, 7) for combinations of all seat sensor observations and the parsimonious combination of these sensors are presented in the Tables 9.4 and 9.5 respectively. Even when all seat movement sensor signals are combined the log-likelihood is considerably lower than for EEG. The high level of correlation and subsequent redundancy between the seat movement sensors is highlighted by the fact that the parsimonious models only require 3 or even 2 (when correlation between observation is taken into account) parameters. The negative signs of all regression coefficients in the parsimonious models confirm the fundamental assumption that reduction in the seat movement magnitude is associated with transition to drowsiness.

For the parsimonious models there is a massive difference of 76.76 in the values of log-likelihood between the seat movement and EEG models (for the case within-transition episode correlation). Finally the ROC curve for one of the parsimonious models for seat movements is presented in Figure 9.3.

Table 9.4 Multivariate binary logistic regression for different combinations of seat movement sensor as covariates and log odds of drowsiness outcome with different correlation assumptions.

Correlation model	preceding values for all sensors (10 parameters)	Latest and preceding values for all sensors (20 parameters)
Independent Observations		
Log-likelihood	−461.129	−452.026
A-ROC	0.673 [0.632; 0.713]	0.682 [0.642; 0.721]
Within same transition episode (random effects)		
Log-likelihood	−454.907	−443.599
A-ROC	−0.668 [0.627; 0.708]	0.677 [0.639; 0.716]
Between all observations for subject (random effects)		
Log-Likelihood	−454.877	−445.583
A-ROC	0.671 [0.630; 0.712]	0.682 [0.643; 0.721]

Key: A-ROC is the area under Relative Operating Characteristic (ROC) curve.

Table 9.5 Parsimonious binary logistic regression models for movement sensor as covariates and log odds of drowsiness outcome with different correlation assumptions.

Correlation model	Covariate 1	Covariate 2	Covariate 3
Independent observations			
Covariate	Seat1_previous_ref	Seat1_recent_ref	Seat5_recent_ref
Regression coeff.	−46.36	−27.55	−70.46
	[−73.36; −19.37]	[−52.66; −2.45]	[−120.26; −20.65]
z	−3.37	−2.15	−2.77
p-value	0.001	0.031	0.006
Log-likelihood		−466.22	
A-ROC		0.682 [0.644; 0.720]	
Within same transition episode (random effects)			
Covariate	Seat1_recent_ref	Seat4_previous_ref	
Regression coeff.	−57.12	−46.24	—
	[−83.55; −30.70]	[−66.61; −25.87]	—
z	−4.24	−4.45	—
p-value	< 0.001	< 0.001	—
Log-likelihood		−459.02	
A-ROC		0.681 [0.643; 0.719]	
Between all observations for subject (random effects)			
Covariate	Seat1_recent_ref	Seat4_previous_ref	
Regression coeff.	−48.30	−40.51	—
	[−69.80; −26.80]	[−58.58; −22.44]	—
z	−4.40	−4.39	—
p-value	< 0.001	< 0.001	—
Log-likelihood		−460.48	
A-ROC		0.681 [0.644; 0.719]	

Key: C4- central site, O2- occipital site, A-ROC is the area under Relative Operating Characteristic (ROC) curve, p-value = significance level, z is the ratio of estimate to its standard deviation.

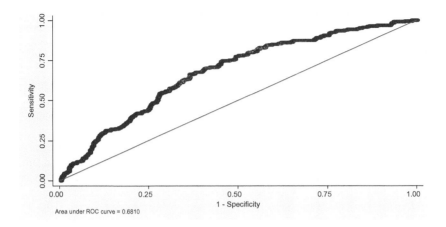

Fig. 9.3 ROC curve for the parsimonious model with the seat movement sensor signals as parameters.

9.3.3   Combination of EEG and Seat Movement Magnitude

The parsimonious models for the combination of EEG and seat movement magnitude are presented in Table 9.6. It is evident that replacing one of the EEG covariates with a seat movement signal significantly increases the log-likelihood. The ROC curve for one of the parsimonious models for a combination of EEG and seat movements' magnitude is presented in Figure 9.4.

9.4 Discussion

The most important findings of the conducted analysis of associations between probability of drowsiness and selected physiological indicators of drowsiness over the course of episodes of transition to drowsiness are:

- Statistically significant associations were established for the analysed physiological indicators – EEG alpha band power percentage and seat movement magnitude. Directions of all associations were physiologically plausible with increase in probability of drowsiness associated with increases in the EEG alpha band power percentage and reduction in the seat movement magnitude;
- The "gold standard" EEG alpha percentage demonstrated the most significant association with the probability of drowsiness and model with the seat movement magnitude has the poorest fit.

Table 9.6. Parsimonious binary logistic regression models for a combination EEG alpha band percentages and movement sensor signals as covariates and log odds of drowsiness outcome with different correlation assumptions.

Correlation model	Covariate 1	Covariate 2	Covariate 3	Covariate 4
Independent observations				
Covariate	C4_recent	O2_recent_ref	O2_previous	Seat1_previous_ref
Regression coeff.	3.96 [2.12; 5.79]	4.54 [2.92; 6.16]	3.71 [2.25; 5.18]	−71.23 [−96.45; −46.00]
z	4.23	5.50	−3.44	−4.40
p-value	< 0.001	< 0.001	0.001	< 0.001
Log-likelihood			−385.69	
A-ROC			0.793 [0.757; 0.829]	
Within same trans. episode (random effects)				
Covariate	C4_recent	O2_recent_ref	O2_previous	Seat1_previous_ref
Regression coeff.	6.03 [3.54; 8.52]	5.50 [3.27; 7.73]	5.46 [3.31; 7.62]	−77.96 [−110.91; −45.02]
z	4.75	4.83	4.97	−4.64
p-value	< 0.001	< 0.001	< 0.001	< 0.001
Log-likelihood			−374.42	
A-ROC			0.790 [0.753; 0.827]	
Between all observations for subject (random effects)				
Covariate	C4_recent	O2_recent_ref	O2_previous	Seat1_previous_ref
Regression coeff.	6.37 [5.80; 10.97]	4.44 [2.48; 6.39]	6.23 [4.38; 8.08]	−71.20 [−99.00; −43.41]
z	6.37	4.45	6.60	−5.02
p-value	< 0.001	< 0.001	< 0.001	< 0.001
Log-likelihood			−361.59	
A-ROC			0.784 [0.746; 0.821]	

Key: C4- central site, O2- occipital site, A-ROC is the area under Relative Operating Characteristic (ROC) curve, p-value = significance level, z is the ratio of estimate to its standard deviation.

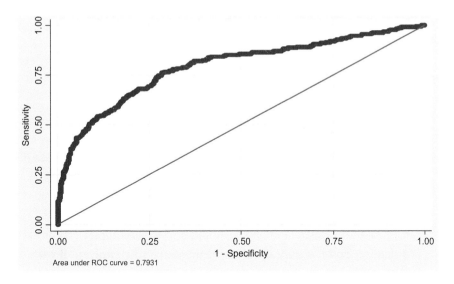

Fig. 9.4 ROC curve for the parsimonious model with EEG alpha percentage and seat movement data as parameters.

- Adding a non-invasive measure that is seat movement magnitude to any parsimonious combination of the EEG derived physiological predictors always resulted in statistically significant improvement of associations;
- The diagnostic accuracy as measured by the area under ROC curve was relatively poor (for an effective practical use of the fitted models) for all combinations of analysed parameters implying the need for more sophisticated derivations of the measured signals. This observation was not unexpected as the signal processing methods used in the analysis were relatively simplistic. The combinations of EEG alpha percentage and seat movement magnitude were found to have largest values of the area under the ROC curve.

This article presented the statistical methodology for analysis of associations between the driver drowsiness and accompanying physiological indicators. The main conclusions are that the non-invasive measures of body movement recorded with piezofilm sensors located inside the car seat demonstrated statistically significant association with the state significant drowsiness and also improved associations with drowsiness when used in combination with other physiological measures namely EEG alpha percentage.

To achieve the practically acceptable level of accuracy of drowsiness detection the detection system could be improved in a number of ways.

- Sensitivity and reliability of measurement of body movement magnitude and spatial patterns should be improved. There are potential benefits in integrating additional sensors at the steering wheel and seat belt;
- Sophisticated signal analysis algorithms should be implemented particularly to deal with the noise and signal artefacts;
- Presented statistical analysis should cover the complete recorded data set and not only episodes of transition to significant drowsiness as conducted at this stage. It is also important to expand the statistical models to the case of multiple categories of drowsiness from the binary model.

However in spite of these deficiencies the authors believe that the presented statistical analysis methodology is the useful starting point that has established associations between drowsiness and non-invasive measures (mainly magnitude of the seat movements) as well as determined the relative effects of various detection techniques and demonstrated benefits of the hybrid approach. With potential improvements in recording accuracy and signal analysis sophistication the described statistical method could still be used to estimate model fit and diagnostic accuracy of more advanced implementations.

9.5 Acknowledgement

We acknowledge support from ARC Linkage grant (LP0347012), Australia.

References

[1] Burton, D., *Vigilance Monitoring system*, in A.P. Office (ed.) no AU 200022710B2, Australia, 2004.
[2] P.P. Caffier, U. Erdmann, and P. Ullsperger, "Experimental evaluation of eye-blink parameters as a drowsiness measure", *Eur J Appl Physiol*, 2003, 89(3–4), pp. 319–325.
[3] A. Craig, K. Hancock, et al. "The lifestyle appraisal questionnaire: A comprehensive assessment of health and stress." *Psychology & Health,* 1996, 11(3): 331–343.

[4] J. Connor, P. Norton, S. Ameratunga, E. Robinson, L. Civil, R. Dunn, J. Bailey, and R. Jackson, "Driver sleepiness and risk of serious injury to car occupants: population based case control study." *Br. Med. J.* 2002, 324 (7346), pp. 1125.

[5] D. F. Dinges, and R. Grace, "PERCLOS: A valid psychophysiological measure of alertness as assessed by psychomotor vigilance", *Federal Highway Administration, Office of Motor Carriers. Report No. FHWA-MCRT-98-006*, 1998.

[6] AJ. Dobson, *An introduction to generalised linear models*. Chapman Hall. 2002.

[7] L. Fletchar and A. Zelinsky, "Context sensitive driver assistance based on gaze – road scene correlation", *International symposium on experimental robotics*. Rio de Janeiro Brazil. 2006.

[8] FB. Hu, J. Goldberg, D. Hedeker et al, "Comparison of population-averaged and subject-specific approaches for analysing repeated binary outcomes", *Am J Epidemiol.* 1998: (147); 694–702.

[9] T. Jung, S. Makeig, M. Stensmo, and T. J. Sejnowski, "Estimating alertness from EEG power spectrum". IEEE Trans. Biomed. Eng. 1997, 44(1), pp. 60–69.

[10] S.K. Lal and A. Craig, "A critical review of the psychophysiology of driver fatigue", *Biol Psychol*, 2001, 55(3), pp. 173–194.

[11] S.K. Lal and A. Craig, "Driver fatigue: electroencephalography and psychological assessment", *Psychophysiology*, 2002, 39(3), pp. 313–321.

[12] S.K. Lal and A. Craig, "Reproducibility of the spectral components of the electroencephalogram during driver fatigue", *Int J Psychophysiol*, 2005, 55(2), pp. 137–143.

[13] K.Y. Liang and S. L. Zeger, "Regression analysis for correlated data", *Annual Review of Public Health* 1993, 14, pp. 43–68.

[14] D. McNair, M. Lorr, et al., *EDITS manual for the profile of mood states,* San Diego, CA, Educational and Industrial Testing Service, 1971.

[15] M. Mahowald, "Eyes wide shut. The dangers of sleepy driving", *Minn.Med.* , 2000, 83(3), pp. 25–30.

[16] P.D. Rochford, W. Ruehland, T. Cherchward and R. J. Pierce, "Evaluation of automated versus manual scoring of polysomnographs on sleep disordered breathing". *Australian Sleep Association Meeting,* 2006.

[17] Santamaria, J. and K.H. Chiappa. *The EEG of Drowsiness*, Demos Publications, New York, 1987.

[18] D. Singer, "Using SAS PROC MIXED to fit multilevel models, hierarchical models and individual growth models", *J Educ Behav Statistics.* 1998 (24): 323–355.

[19] Spielberger, C. D., R. L. Gorsuch, et al, *Manual for the state-trait anxiety inventory*, Palo Alto, CA, Consulting Psychologists Press, 1983.

[20] J. A. Stern, D. Boyer, and D. Schroeder, "Blink rate: A possible measure of fatigue", *Human Factors*, 1994, 36(2), pp. 285–297.

[21] Verbeke, G. and G. Molenberghs , *Linear models for longitudinal data*. Springer-Verlag. New York. 2000.

[22] W.W. Wierwille and L.A. Ellsworth, "Evaluation of driver drowsiness by trained raters", *Accident Analysis & Prevention*, 1994, 26(5), pp. 571–581.

[23] S. Zeger and KY. Liang, "An overview of methods for the analysis of longitudinal data", *Stat Med.* 1992: (11); 1825–1829.

[24] E. Zilberg, ZM. Xu, D. Burton, M. Karrar, and S.K. Lal, "Statistical validation of physiological indicators for non-invasive and hybrid driver drowsiness detection system", AusWireless2007, Australia, 2007.

[25] E. Zilberg, "Analysis of associations between expert rating and physiological indicators of driver drowsiness", Master of Biostatistics Thesis. University of Melbourne, Australia, 2007.

10

Security and Privacy of Wireless Sensor Networks (WSN) for Biomedical Applications

Ellen Stuart, Teng-Sheng Moh and Melody Moh

San Jose State University, USA

Wireless sensor network applications in healthcare and biomedical technology have received increasing attention, while associated security and privacy issues remain open areas of consideration. The relevance of this technology to our growing elderly population, as well as our increasingly over-crowded and attention-drained healthcare systems, is promising. However, prior to the emergence of these systems as a ubiquitous technology, healthcare providers and regulatory agencies must determine an acceptable level of security and privacy. This chapter will provide an introduction to biomedical applications of wireless sensor networks, identify security and privacy issues to be addressed, and survey some of the proposed methods for securing these systems.

As communication technologies advance and become more diverse, biomedical applications of wireless sensor networks (WSN) can be generally categorized in one of three areas:

- Patient habitat monitoring systems
- Patient monitoring systems (monitoring of biological systems)
- Actuator-enabled devices that affect a functioning system

This chapter will review these areas of application in more detail, describing examples of systems that have been proposed or implemented. Along with these examples, we identify the following privacy and security obstacles associated with biomedical applications:

- Limited resources
- Fault tolerance, interference, and attacks

- Confidentiality
- Physical security

While some of these security risks are common to many wireless and general networking protocols, we identify how these risks are specifically relevant to wireless sensor networks resulting in the necessity of security solutions that extend to fit the additional constraints of these specialized biomedical systems. As we will describe, the groundwork for these solutions exists in models, requirements, and solutions designed to address the obstacles presented by these innovative applications.

This chapter is organized as follows: Section 10.1 is a review of fundamental properties of a wireless sensor networks and components of these systems. In Section 10.2 we will describe the areas of biomedical applications of WSN applications in more detail. Section 10.3 identifies security issues that are relevant to wireless sensor networks and specifies what specific risks might be relevant to biomedical applications of this technology. Section 10.4 is a survey of approaches to securing wireless sensor networks and a discussion of the relevance to one biomedical applications of this technology. Finally, Section 10.5 concludes with the identification of future directions of research and development of security solutions for WSNs, supporting systems, and related fields.

10.1   Wireless Sensor Networks

A wireless sensor network consists of distributed autonomous nodes or motes that are enabled by sensors. These devices can used to monitor their surrounding environment, including temperature, motion, sound, and pressure. While these wireless sensor nodes or motes are typically organized similarly to mobile ad-hoc networks (MANET), they are typically implemented using extremely compact devices with limited processing capabilities, limited memory, and operating with limited battery power.

These constraints go beyond the limits of MANET systems and mandate that specialized protocols provide for communications within the WSN. Power limitation results in further constraints on the number and distances that motes can transmit, and WSN's typically employ some type of intermediary base station or relay device within the system architecture to provide for more energy intensive activities. These activities would include node authentication, key distribution, and the transmission of raw or aggregated data via the internet.

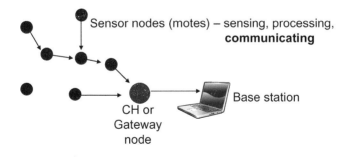

Fig. 10.1 Basic Wireless Sensor Network design.

10.2 Biomedical Applications of Wireless Sensor Networks

A study by Plunkett Research points out that the rapidly aging population of the U.S. population and the increase life expectancy of seniors will result in a significant strain on the health care system [42]. While Plunkett indicates the "total U.S. health care expenditures are projected to increase from $2.39 trillion in 2008 to $2.72 trillion in 2010," 78 million Baby Boomers who begin turning 65 in 2011 could cause the burden on government and health care providers to grow and balloon as this population continues to age.

With current annual nursing home and home health costs nationwide totaling approximately $198.5 billion [42], possible applications of WSN that would automate some amount of this care or make it more efficient are very relevant in addressing this potential catastrophe.

Implementations of WSN in healthcare usually involve sensors that are placed in close proximity to a person, on the body (extracorporeal), or inside the body (in-vitro). In-vitro devices are typically implanted or ingested, depending on the desired positioning. These sensor nodes form a network and in certain applications, can be referred to as a body area network — BAN, or personal area network — PAN. As in more traditional WSN applications, the motes either forward data to other sensor nodes as relays, or to a cluster head device that can relay information to a base station, located nearby.

At this point in WSN development, the specific implementation of the application determines how data is aggregated, analyzed, forwarded, and secured. Surveying the broad range of proposed and implemented applications of WSN, systems typically fall into one of the following categories described in the remainder of this section.

Patient Habitat Monitoring

Similar to traditional habitat monitoring applications that utilize WSN technology [21], patient habitat monitoring systems provide information regarding the patient's activity and/or environment [5, 9, 10, 16, 28, 30]. Within biomedical monitoring applications that have been proposed and implemented, there is a wide range of architectures and approaches.

One of the more specialized systems utilizes radio frequency identification (RFID) and WSN motes to schedule and monitor patient medications dispersal [10]. Other applications that support remote healthcare systems seek to track environmental information and find any correlations between patient condition and external influences such as temperature, UV exposure, atmospheric condition, pollen, etc. [28]. Many of these designs are targeted towards creating a cost effective remote healthcare solution in the wake of a growing elderly population and rising healthcare costs.

Monitoring of Patient Condition

Aside from habitat monitoring, the potential value of WSN in the field of healthcare and biomedical applications is in the monitoring of physiologic parameters whereby streaming data can be analyzed and recorded over an extended period of time. This real-time continuous monitoring automates the

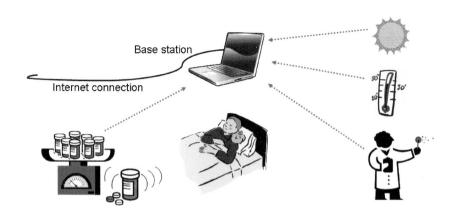

Fig. 10.2 Patient Monitoring.

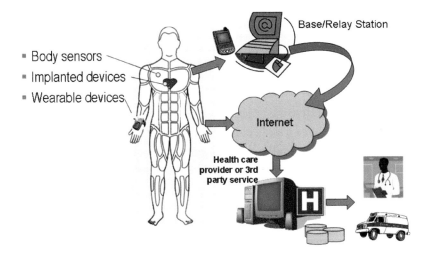

Fig. 10.3 Patient Monitoring Systems.

analysis and alert of conditions requiring physician care [9]. This can be especially effective in the treatment of chronic illnesses such as diabetes or heart conditions, and could be combined with automated healthcare delivery and emergency medical services (Figure 10.3).

Proposed or implemented patient monitoring applications include:

- Monitoring of vital signs — measuring and tracking of vital signs, including body temperature, ECG activity and heart rate, blood pressure, and blood oxygen levels [5, 8, 28, 30]
- Stress monitor — using sensors to establish baseline heart rate variability (HRV), followed by sampling in extended monitoring periods [15]
- Blood glucose and insulin pump monitor [5, 15, 30]
- Organ monitor — using sensors to monitor conditions during organ transplant including in vitro heart carbon dioxide, oxygen, and methane to determine heart viability during transplant [30].
- Cancer detector — proposed monitoring of blood flow changes/anomalies due to nitric oxide levels (cancer cells studies shown to exude nitric oxide and affect the blood flow around a tumor); also, proposals for possible sensor placement on needles to provide for diagnosis without biopsy [30].

- Action recognition — system using on-body microphones and sensors (three axis accelerometers), information is gathered and transmitted for further analysis and activity identification [38].
- Personal Locator — using positioning system [15].

Wang et al summarize the potential value of these applications in providing patient histories to create a more complete medical record [37]. Wang identifies the importance of these applications in providing capabilities beyond a data repository; that is, enabling anomaly detection, identification of drug side effects, and the identification of lifestyle trends or activity patterns.

Hybrid solution

Riudavets proposes an application that combines patient habitat and patient vital sign monitoring [28], providing a remote care monitoring system. This is one of many solutions that utilize the Multi-Router Traffic Grapher (MRTG) tool to collect and analyze results. In this system, MRTG facilitates the determination of correlations between the patient's environment and their health conditions as indicated by recorded vital signs. In addition to providing real-time monitoring of patient condition, systems like this can provide a graphical representation of patient case history (in daily, weekly, monthly or yearly

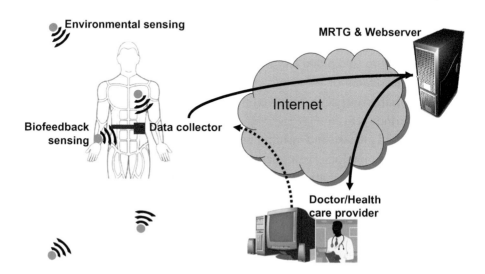

Fig. 10.4 Hybrid Solution — Correlating patient and patient habitat monitoring.

graphs) and make objective information available to researchers and emergency or regular health care providers.

Actuator-Enabled WSN Applications

Actuator-enabled wireless sensor network systems are a highly specialized group of WSN applications that integrate WSN technology with advanced actuator technology. Examples of these applications include the development of an artificial retina or retina prosthesis chip utilizing sensor networks [30], the integration of WSN with neural networks to train robotic action [24], and the combination of WSN technology with stimulators to develop more advanced brain activity regulators/stimulators [39] or functional electrical stimulator (FES) systems.

Jones describes deep brain sensors and stimulators that have been used to treat Parkinson's disease, epilepsy, intractable seizure and other movement disorders [14]. Jones also notes future applications that could utilize advanced microelectronics, micro-devices, and bio-nanotechnology to provide for metabolism control, energy conversion, cell analysis, blood diagnosis, drug/DNA delivery systems, hormones as carrier molecules, molecule detection (e.g. allergens or viruses), physical check-ups, and artificial control of immunoreactions. Also, while only an application in animals, WSN have been proposed and implemented for real-time autonomous animal control [39].

From this spectrum of applications emerges an equally broad array of security and privacy concerns which will be identified in the next section.

10.3 Privacy and Security Issues

WSN are subject to many of the same attacks as wireless communications in general, but face additional limitations due to environment, resources, and their potentially dynamic and distributed nature. Biomedical applications of this technology are especially vulnerable due to the sensitivity of personal and patient data, the unique, possibly implanted, environment, and potential volatility of the ensuing action (life threatening). In this section, some of these obstacles will be described in more detail, as well as their specific relevance to the biomedical applications previously identified.

Hu *et al.* describe a secure WSN solution as inclusive of confidentiality, authenticity, integrity, and freshness [13]. Kulkarni extends the security requirements for WSN systems in healthcare as providing secure communication links with healthcare providers, robust network security, secure sensing and monitoring devices, and strong patient-provider authentication [16].

In more traditional sensor network monitoring solutions, systems are characterized by a hostile or uncontrolled environment, limited battery, processor, memory, bandwidth, transmission range, and a dynamic and distributed architecture. While biomedical applications of WSN are also subject to the same limited resources and dynamic nature, they typically do not operate in the same type of hostile or uncontrolled environment. However, solutions for these systems must address the more novel environment of the patient surrounding or the body itself. These obstacles must be accounted for in a WSN security solution, giving thorough consideration for the vulnerabilities of biomedical applications, discussed in the rest of this section.

Limited Resources

While WSN protocols are in some ways constrained by a dynamic nature, similar to that of an ad hoc mobile network, the reduced size of the WSN motes places more extreme constraints on the protocols that can be considered feasible. This includes the mechanisms used for the computation of encryption protocols, key length, key and data storage, and energy supply. The limited energy supply affects all of the fundamental mote activities including idle "listening", sensing, processing, and transmission activities which are especially energy intensive.

Although energy and processing resources must be considered in the design of any security protocol, this is a critical factor in biomedical applications, where device lifespan could affect the well-being of a host/patient. In systems where the sensor nodes are implanted within the body, device malfunction or security failures due to power loss or an ineffective security protocol can be fatal. The majority of WSN security mechanisms are designed and evaluated for energy efficiency, trying to maximize sensor node lifetime under the constraint of extremely limited power supply.

Fault Tolerance, Interference, and Attacks

In WSN systems, fault tolerance must be considered in the context of battery depletion, transmission failure (either temporary or persistent), routing, attack, and congestion. A secure biomedical application must have measures in place for continuous operations and safeguards to prevent, detect, isolate, identify, and recover from system failures and attacks [27]. It is also important to provide a fault tolerant solution that prevents transmission issues from blocking network function with unnecessary congestion, which could also be exploited in denial of service attack, as well as cases of disruptive interference involving WSN attributed to:

- Multiple subjects/WSNs, resulting in conflicting data
- Interference to/from other devices
- Denial of Service (DoS) attacks, resulting in non-functioning nodes/clusters/networks, loss of power, transmission or aggregation of conflicting/erroneous information

Any interference from competing devices, networks, or environmental factors could be misinterpreted, and the ensuing actions could be extremely harmful or disruptive. Similarly, medical devices can be affected by the electromagnetic interference generated by WSN systems. In cases where signals are mistaken for physiological conditions that are being monitored by the devices, this could result in physical harm and misdiagnosis [4].

One fundamental difference between WSNs and other types of more traditional networks is the physical vulnerability of the distributed network. Because of the redundancy within this architecture, a network's distributed messages, routing communications, security keys, and devices themselves can all be vulnerable to the compromise of a single node [3]. A compromised node could result not only in a lack of coverage, but also may be subject to analysis, cloning, and subsequent attack based on the captured information.

Lee and Choi identify that once a node is compromised, it may be repeatedly replicated/cloned. The attacker can utilize this to establish new relationships with neighboring nodes within the network, eavesdrop using cryptographic information taken from the compromised node, inject false reports, or selectively drop message packets without forwarding [18].

Anand describes three categories of attacks [3]:

- Eavesdropping — collecting data from communications, either in an indirect manner (listening to communication) or direct method of querying network components
- Disruption — altering the data that is communicated by the sensor, either by submitting false data or by affecting a legitimate sensor to corrupt its report

- Hijacking — utilizing a captured legitimate node(s) to spread false data or drop valid data

Jamming is another attack that would exploit both the network's vulnerability to interference, as well as the limited power of the networks elements. Law describes jamming attacks and techniques, including attempts to create intentional interference, as well as attempts to create superfluous RTS packets ([14] describes "Hello" attack) that effectively prevent legitimate requests from being served [17]. Law defines and compares the effectiveness resource requirements of the following categories of jamming techniques:

- Constant jammer — continuously jams legitimate communications by creating constant noise as interference
- Deceptive jammer — fabricates or replays valid signals to introduce spurious data into the network communications; prevents meaningful information from being exchanged or validated
- Random jammer — uses sleep/wake cycle of jamming to create a less effective, but more energy efficiency attack (compared to the constant jammer)
- Reactive jammer — listens for activity on the channel, and in case of activity, immediately sends out a random signal to collide with the existing signal on the channel

Law also identifies energy efficient jamming techniques that increase effectiveness of jamming on protocols where timed communication periods associated with data exchange can be identified, and attacks can be more successfully targeted. These DoS attacks must be addressed through a fault tolerant security scheme that can detect and mitigate attacks.

Confidentiality

Managing and authenticating communications in a dynamic wireless network can be challenging, and is magnified in wireless sensor nodes where the device landscape is dynamic and resources are limited. Because of the data being transmitted in the systems described above, confidentiality must be constant and without question. This confidentiality needs to be provided at all transmission points between sensor, forwarding or aggregation activities, and base station.

Distributed encryption keys systems must be planned with all of the vulnerabilities of the system in consideration. The capture of a single sensor node could render an entire shared key system useless. Encryption schemes are generally classified as symmetric key or public keys systems, or some combination of the two. Symmetric key systems, typically requiring less overhead [23], are sometimes cited as being more feasible to the limited energy and processing capabilities of the nodes used with WSN. However, public key encryption systems might more feasibly support the dynamic nature of the WSN. Encryption schemes must be sophisticated enough to ensure sensitive data is protected. This must be accomplished with keys of limited size on processors with established limits.

Physically Security

Physical security is an issue that can be partially addressed by providing for the obstacles identified above. Additionally, because many of these solutions require direct contact or implantation of devices, biomedical applications should be especially safe and reliable.

As wireless technologies, including WSN, radio frequency identification (RFID), WPAN, and telemedicine, become more and more pervasive, the resulting ecosystem becomes more complex. It becomes obvious that the safety issues and side effects of these technologies must be thoroughly considered prior to use with human hosts. These factors must be integrated into the design of such applications, and of the healthcare environment as a whole. It is vital that possible sources of interference be thoroughly considered and addressed in the WSN system so that external signals are not either mistakenly interpreted as a physiological event, existing condition, or interfere, mask, or block intentional sensor function.

The vulnerabilities presented by the proximity of these WSN devices to sensitive organs, as well as the growing intricacy of wireless communication and powered medical devices used in the healthcare systems, make interference a growing concern. Recent studies and FDA reports have identified cases where interference caused by coexisting RFID devices has resulted in device malfunction [4, 35, 36]. In one case the interaction resulted in the reprogramming and temporary malfunction of a wireless deep brain stimulator, inducing a severe rebound tremor in the patient [35]. Ashar and Ferriter identify a possibility that similar scenarios could occur with WSN, with an increasing potential for continued incidents as the transmission space continues to become more congested [4]. In cases where sensors are implanted in tissue, it is also important to consider the more direct effects of energy absorption variability in the implant environment [30].

10.4 Security Solutions

WSN, like other wireless communication systems (and any other computer network), are implemented with some degree of vulnerability. Because of their limited resources, dynamic nature, and operating environment, the development of a secure WSN communication scheme is impeded by numerous obstacles. While some of these challenges are present with other systems, we've described how the nature of both the WSN and the unique operating environments of biomedical applications often present novel scenarios and possible personal risk. In this section, we describe proposed solutions that address specific vulnerabilities of WSN operating in the biomedical environments indicated in the previous sections. While these proposals address key elements of a secure system, no complete solution exists for WSN or for biomedical-specific applications of this technology.

Limited Resources

Many research studies and written submissions on the topic of WSN security are focused on the development of an energy efficient means to provide security while still providing some level of security and confidentiality. To determine the suitability of the encryption algorithms for use with biomedical WSN, Strydis et al study various symmetric key block ciphers, comparing them based on power consumption, energy expenditure, encryption rate, executable size, and security margin, and determined that the MISTY1 algorithm showed the most promising results across their metrics [31] (See Table 10.1).

Other protocols such as the Hybrid Indirect Transmission (HIT), proposed by Culpepper et al, are structured to reduce transmission cost and extend network lifetime through the use of clustering and data fusion [6]. This specific method utilizes these established mechanisms, along with common medium access protocols such as Carrier Sense Medium Access with Collision Detection (CSMA/CD) and Time Division Medium Access (TDMA) to provide for a dynamic and energy efficient system. Data is communicated within and among clusters using parallel transmissions.

Table 10.1 Results of block cipher comparison by Strydis.

Average Power Consumption	Peak Power Consumption	Total Energy Cost	Energy Efficiency	Encryption Rate	Program-Code Size
IDEA	IDEA	RC6	RC6	RC6	XXTEA
LOKI91	MISTY1	RC5	IDEA	RC5	3WAY
SKIPJACK	LOKI91	IDEA	RC5	MISTY1	LOKI91
MISTY1	TWOFISH	MISTY1	MISTY1	RIJNDAEL	RC6
RIJNDAEL	RIJNDAEL	BLOWFISH	RIJNDAEL	BLOWFISH	RC5

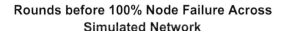

Rounds before 100% Node Failure Across Simulated Network

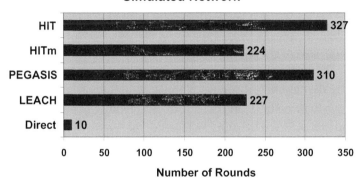

Fig. 10.5 Comparison of simulated network lifetime for HIT protocols.

Culpepper introduces single cluster (HIT) and multiple cluster (HIT$_m$) versions of the protocol, and compares these to the network lifetime of direct transmission and two more established WSN protocols, LEACH and PEGASIS.

HIT is designed to operate in rounds, comprised of the following periods:

• Cluster-Head Election — new cluster-heads (CH) are elected
• Cluster-Head Advertisement — each CH broadcasts its status to the network
• Cluster Setup — Clusters and the upstream and downstream relationship are formed

- Blocking Set Computation — Each node computes its blocking set (prevent simultaneous transmission/interference during between nearby nodes)
- Route Setup — Sensors within a cluster form multi-hop routes to the cluster-head
- MAC Schedule Creation — A TDMA schedule is computed to allow for parallel transmissions, each sensor independently computing a medium access controlling TDMA schedule
- Data Transmission — A long steady state transmission phase where sensed data is communicated to the base station using TDMA

In a modified version (HITm) limits were set, used to prevent the re-election of possible attackers who attempt to railroad the system by exploiting self-election and priorities based on self-identification.

Culpepper evaluates HIT and HITm against LEACH, PEGASIS, and Direct Transmission, based on energy dissipation metrics calculated as the product of the average energy and delay product. These energy values are averaged by the number of rounds completed before all nodes had depleted their energy (Figure 10.5). HIT was shown to require about 25% of the time required by PEGASIS or LEACH, in terms of the length of time for each round of data gathering

Culpepper also showed that as the network scaled, the improved performance of both HIT and HITm became more significant. While HIT does seem to provide energy savings and reduced delay over LEACH and PEGASIS, it proposes the use of data fusion to increase the efficiency of data communications throughout the network. The feasibility of a secure solution utilizing data fusion must be considered and evaluated in the context of a proposed application.

In another study, Zia et al introduces and evaluates the TripleKeys security mechanism against two other protocols that address cluster based hierarchical WSNs: TinySec and MiniSec [41].

The 'triple keys' in the protocol name refers to the following:

- K_n (network key) — Key used by nodes to encrypt data and forward to next hop
- K_s (sensor key) — Key used by the cluster leader for encryption and decryption of communications with the base station

- K_c (cluster key) — Key generated by the cluster leader, and shared by the nodes in that particular cluster for encryption and decryption of data to and from cluster leader

TripleKeys requires that each node have mutual authentication with its neighbors and cluster leaders, in order to provide a level of protection against sinkhole and Sybil style attacks. Zia evaluates the results based on the resources required for execution. TripleKeys was shown to have the lowest overhead and also overcomes the security and routing weaknesses identified in TinySec (TinySec does not provide a secure localization or secure routing mechanism). Zia noted that although MiniSec provides a good level of security, the required overhead is much more than both TinySec and TripleKeys, making it less suitable for WSN application.

Confidentiality

As noted in the above section with regard to the HIT protocol, data aggregation method for sensor data from multiple sensory nodes (possibly in distributed manner) needs to provide for security and confidentiality in biomedical applications of WSN. Data aggregation techniques must consider fault tolerance, expansion of capacity, and robustness [20]. Because biomedical applications must provide both confidentiality and accessibility, these must also be preserved through both data communication and aggregation schemes.

As with other computer and networking systems, confidentiality can be addressed, in part, through the use of an effective authentication and encryption system. Although some systems utilize a centralize authority (such as Kerberos) to provide authentication, wireless sensor nodes cannot realistically be authenticated through a remote authority. In some BM WSN systems, this is mimicked using a base station that is in close proximity to more resource intensive processes, including provide authorization.

Another approach, key pre-distribution is cited as the most common method of key establishment in WSNs due to constrained resources and much of sensor network security is based on this practice [12, 23]. While this key pre-distribution resolves the challenge of issuing keys, it is also important to consider and prevent the compromise of a single node (and its key) from affecting the entire network by ensuring that motes are either tamper resistant or are assigned varying symmetric keys per mote.

Oliveira et al propose an enhancement (SecLEACH) to a previous design (LEACH) using a method of random key pre-distribution to secure it [25]. Hu et al introduce a solution targeting the specific framework and application for WSAN using multiple keys and cite that while keys can be established prior to release, re-keying is crucial in preventing the capture of permanent keys with the compromise of a single node and also enables a dynamic architecture [12].

Liu and Ning describe a solution for key pre-distribution assuming a known, approximate architecture (predetermined locations) [19]. Using a planned arrangement of expected neighbors, sensor nodes are provided with pre-selected polynomials to be used for key generation. While this might be too rigid as a general approach for WSN security, it might be relevant to specific applications with a planned architecture, such as a healthcare facility or an implanted sensors system.

An alternative to symmetric key systems, asymmetric or public key cryptography is sometimes identified as infeasible given the limited resources of a mote. Public key systems typically require increased processing, memory, and energy resources, but might be more appropriate for the dynamic architecture of WSNs. Many of the public key systems that are proposed for WSN applications utilize elliptical curve cryptography (ECC), a public key encryption algorithm that can provide effective level of security (relative to other public key systems) with a smaller key size [7, 22, 34]. Table 10.2 shows a comparison of the equivalent effectiveness of key size between an ECC algorithm, symmetric keys system, and another asymmetric key system (RSA).

Implementations of public key encryption in sensor networks is frequently proposed using Elliptical Curve Cryptography (ECC) to address resource

Table 10.2 Equivalent key sizes for symmetric, ECC and RSA.

Sym key (bits)	ECC key (bits)	RSA key (bits)
80	160	1024
112	224	2048
128	256	3072
192	384	7680
256	512	15,360

constraints [7, 22, 34]. Malasri and Wang present a solution, Sensor Network for Assessment of Patients (SNAP), designed to address privacy, security, and integrity of data in medical sensor networks [22]. They define a process that involves multiple layers of authentication and the update of encryption keys, using multiple encryption schemes (including ECC) to provide secure communications

A more unusual authentication system, proposed by Falck is designed specifically for use with body area networks (BANs) [8]. Body-coupled communication (BCC) is used to allow sensors to identify other sensors associated with the same body, but not create interference with other systems. A unique identification is created by the generation of a weak electrical field that extends only centimeters beyond the skin and serves to authenticate related sensors within the same BAN.

Fault Tolerance, Interference, and Attacks

As noted in the previous section, fault tolerance must provide for failures due to depletion of battery, transmission failure (either temporary or persistent), destruction from non-intentional event, and congestion within network. Paradis and Han describe methods for addressing each of these areas, but also note that no existing protocol can provide for complete fault tolerance [27]. Their methods include prevention (typically through routing redundancy), detection and recovery (normal performed in network layer protocols), and isolation and identification.

One of the methods designed to prevent interruption or attack is the Neighbor Watch System, which uses Neighbor List Verification and redundant communications to prevent communication failures when some part of the system is compromised [18]. In this system, sensor networks using multi-hop routing keep track of the communications of neighboring nodes. The neighboring and transmitting nodes listen and buffer data until the recipient node correctly forwards data to a valid receiver. If it does not, the neighboring and transmitting node(s) forward the message. Because transmission is by far the most costly process for motes, the energy costs of redundant transmissions as messages are forwarded along multiple paths makes this system less than optimal.

Physical attack is another vulnerability that is a risk factor for distributed motes and must be taken into account. Panja et al propose a system of role based security which, among other features, is aimed toward the mitigation of physical attack [26]. Panja uses various security levels and group keys to prevent a single compromised node from impeding other functions across the network.

Hsieh et al introduce Secure Communications of Cluster-Based Sensor Network (SecCBSN), another process intended to minimize the effects of physical compromise to a sensor node [11]. In this cluster-based system, the BS acts as certificate authority, and assigns individual key chains, a function, an individual key, and a certificate (TCert) before deploying the node. The security system is comprised of three modules: primary security, intrusion detection, and cluster round.

With the primary security module, existing nodes can authenticate new incoming nodes, establish secure links, and broadcast authentication between neighboring nodes. The motes use their issued TCert to complete an authentication exchange with neighboring node(s) and establishing pairwise keys with each of the neighbor nodes which then broadcast a successful authentication.

The intrusion detection module contains alarm return, trust evaluation, black/white lists schemes. In SecCBSN, nodes follow a transmission monitor schedule broadcast by their cluster head (CH). Alarm return protocols enable monitor nodes to send alarm packets to the BS, triggering the evaluation or re-evaluation of trust values for 'accused' nodes. The BS evaluates sensor nodes with a trust value and records this value to either a blacklist or white list. Blacklist nodes are restricted from the network and white list nodes are candidate member nodes (MN).

During the cluster round, cluster heads are self-determined and non-CHs join a cluster. The BS can provide each CH with a list of white list nearby. Candidate nodes with higher trust values have higher priorities to join the cluster.

Another attack that was detailed in the previous section is jamming. While there are no adequate counter measures to jamming or persistent interference, Ahmed et al propose a solution to identify routing issues, node malfunction, battery depletion, and jamming (either intentional or incidental). Upon detection, a high priority 'JAM' signal is broadcast to identify its status [2]. This is effective for single channel motes devices but is only effective if the jamming attack is intermittent.

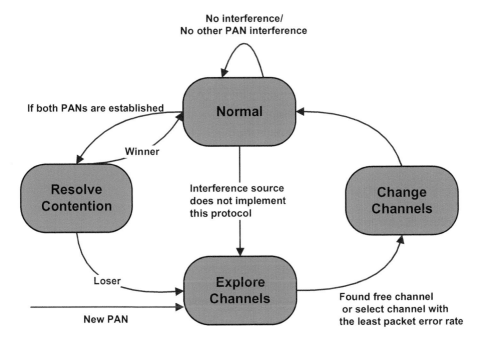

Fig. 10.6 Shah and Nachman's multi-channel solution.

For multi-channel motes, frequency hopping provides some alternative countermeasure. Shah and Nachman also propose a solution for BAN where interference is detected by a 'coordinator' (a sort of relay station) that checks for other competing packets [29]. Based on a system of priorities, the lower priority entity will search for an alternate channel.

Physical Security

In the previous section, we identified the importance of identifying and addressing the dangers of interference between medical devices and other systems (RFID, WSN, etc.). In biomedical applications, especially systems placed directly in contact with the body, we must also address the dangers that are present and contained within the proposed WSN system. Several routing algorithms have been proposed and tested to address possible affects to tissue surrounding an implanted device [33]. These thermal aware routing algorithms attempt to alter routing based on the temperature effects of sensor node and transceiver operations. In the following protocols, routing paths are determined by various factors, from shortest possible hop count, neighbor node temperatures, and total route temperature increase:

- Thermal-aware routing algorithm (TARA)
- Least temperature routing (LTR)
- Adaptive least temperature routing (ALTR)
- Least total-route-temperature (LTRT)

10.5 Conclusion

Future Directions

In order to provide for the rapid advancement of WSN technology, Sugihara and Gupta suggest the development of a more sophisticated programming model, more suitable for addressing the energy, scalability, fault-tolerance, and collaboration requirements necessary [32]. Ács et al, citing the broad range of applications and numerous studies that offer security solutions to specific scenarios or specific weaknesses, propose a formal security model for WSN [1]. This is suggested as a means, not only for proving a network secure, but also for identifying the weakness in proposed solutions.

Research and development of WSN applications within the healthcare industry continue to become more refined, but a more complete security mechanism is needed to address existing weakness in security. Martin notes the development of more sophisticated jamming identification methods, including functionality for directional analysis and prediction of movement around a WSN [23].

Innovation and development continues for monitoring applications and more advanced actuator enabled bionic systems. Also solutions might spread outside of the healthcare environment, to neighboring industries such as sports and fitness training. The spread of technology necessitate the creation of an infrastructure or platform to provide support for varying biomedical applications including possible publish and subscriber service for collected data [9].

Because of the large amounts of raw data, processing, storage, and analysis requires that advance algorithms are needed to filter and identify useful information. Neural networks are a pattern recognition tool that might be utilized to provide real-time analysis of sensor data.

Summary of Chapter

In this chapter we have surveyed biomedical applications of sensor network technology. Because of the direct link that potentially exists to both personal information and to a person and their welfare, it is especially important to continue re-evaluating privacy and security issues within these systems and their surrounding environments. This becomes vital as the wireless ecosystem becomes more and more dense.

We identified the lack of a complete security solution for biomedical applications of WSN, and yet, implementations of this technology exist. As in other forms of wireless communication, identifying the acceptable level of risk becomes part of the development process, and in this case, must work in conjunction with subject matter experts (healthcare providers) and regulatory agencies (e.g. FDA). Network security is an evolving matter, but before WSN technology can fully emerge as a ubiquitous solution that interacts with existing systems and practices, security and privacy must be addressed in a holistic manner. The existing solutions and proposals will continued to be developed, refined, and implemented into the next generations of wireless sensor network technology, offering promise to the growing demands on the healthcare industry.

Problems

Problem 10.1

Draw a typical biomedical implementation of a wireless sensor network patient monitoring system, including motes, data provider, base stations/gateway devices, and any communication channels.

Problem 10.2

Identify the qualities of a sensor node that put limitations on available security solutions.

Problem 10.3

What is the MRTG (Multi Router Traffic Grapher) tool used for?

Problem 10.4

What methods can be used to address jamming attacks?

Problem 10.5

Describe different types of attacks on wireless sensor networks.

Problem 10.6

When considering encryption of data in a wireless sensor network, what are some of the key systems used?

Problem 10.7

What are some issues when considering the physical safety of a biomedical application of wireless sensor network?

Problem 10.8

Describe the HIT protocol and its mechanisms to reduce transmission cost and overall energy use.

Problem 10.9

Identify the 7 periods the comprise HIT and the associated MAC protocol.

Problem 10.10

Describe the Neighbor Watch System protocol and describe the trade-off between fault tolerance and energy efficiency.

Problem 10.11

Describe routing methods that address physical security/safety issues that are relevant to implanted sensor nodes.

References

[1] G Ács, L Buttyán, I Vajda, Modeling adversaries and security objectives for routing protocols in wireless sensor networks, Proceedings of the 4th ACM workshop on Security of ad hoc and sensor networks, Alexandria, VA, pp. 49–58, 2006.
[2] N Ahmed, S Kanhere, S Jha, The holes problem in wireless sensor networks: a survey, ACM SIGMOBILE Mobile Computing and Communications Review, V. 9, i. 2 (Apr 2005), p. 4–18.
[3] M Anand, E Cronin, M Sherr, M Blaze, Z Ives, I Lee, Sensor network security: more interesting than you think, Proceedings of the 1st conference on USENIX Workshop on Hot Topics in Security, V. 1, Vancouver, B.C., pp. 5–5, 2006.
[4] B Ashar, A Ferriter, Radiofrequency Identification Technology in Health Care:Benefits and Potential Risks, JAMA, Nov 21, 2007; 298: 2305–2307.
[5] W Baumann, M Lehmann, A Schwinde, R Ehret, M Brischwein, B Wolf, Microelectronic sensor system for microphysiological application on living cells, Sensors and Actuators B. v55. 175–180.

[6] B Culpepper, L Dung, and M Moh, Design and analysis of Hybrid Indirect Transmissions (HIT) for Data Gathering in Wireless Micro Sensor Networks, ACM Mobile Computing and Communications Review (MC^2 R/), pp. 61–83, Jan/Feb 2004.

[7] B Doyle, S Bell, A Smeaton, K Mccusker, N O'Connor, Security Considerations and Key Negotiation Techniques for Power Constrained Sensor Networks, The Computer Journal, V. 49, i. 4 (Jul 2006), pp. 443–453.

[8] T Falck, H Baldus, J Espina, K Klabunde, Plug 'n play simplicity for wireless medical body sensors, Mobile Networks and Applications, V. 12, i. 2–3 (Mar 2007), pp. 143–153.

[9] M Gaynor, S Moulton, M Welsh, E LaCombe, A Rowan, J Wynne, Integrating Wireless Sensor Networks with the Grid, IEEE Internet Computing, V. 8, no. 4, pp. 32–39, Jul/Aug, 2004.

[10] L Ho, M Moh, Z Walker, T Hamada, C Su, A prototype on RFID and sensor networks for elder healthcare: progress report, Proceeding of the 2005 ACM SIGCOMM workshop on Experimental approaches to wireless network design and analysis, Aug 22–22, 2005, Philadelphia, PA.

[11] M Hsieh, Y Huang, H Chao, Adaptive security design with malicious node detection in cluster-based sensor networks, Computer Communications, V. 30, i. 11–12 (Sept 2007), pp. 2385–2400.

[12] F Hu, W Siddiqui, K Sankar, Scalable security in wireless sensor and actuator networks (WSANs): integration re-keying with routing, Computer Networks: The International Journal of Computer and Telecommunications Networking, V. 51, i. 1 (Jan 2007), pp. 285–308.

[13] N Hu, R Smith, P Bradford, Security for fixed sensor network, Proceedings of the 42nd annual Southeast regional conference, Huntsville, AL, pp. 212–213, 2004.

[14] V Jones, From BAN to AmI-BAN: micro and nano technologies in future Body Area Networks. Proceedings Symposium on Integrated Micro Nano Systems, 20 Jun 2006, Enschede. pp. 22–29.

[15] E Jovanov, A Lords, D Raskovic, P Cox, R Adhami, F Andrasik, Stress monitoring using a distributed wireless intelligent sensor system, IEEE Eng. Med. Biol. Mag., v. 22, no. 3, pp. 49–55, May/Jun 2003.

[16] P Kulkarni, Y Öztürk, Requirements and design spaces of mobile medical care, ACM SIGMOBILE Mobile Computing and Communications Review, V. 11, i. 3 (Jul 2007), pp. 12–30.

[17] Y Law, L van Hoesel, J Doumen, P Hartel, P Havinga, Energy-efficient link-layer jamming attacks against wireless sensor network MAC protocols, Proceedings of the 3rd ACM workshop on Security of ad hoc and sensor networks, Alexandria, VA, pp. 76–88, 2005.

[18] S Lee, Y Choi, A resilient packet-forwarding scheme against maliciously packet-dropping nodes in sensor networks, Proceedings of the 4th ACM workshop on Security of ad hoc and sensor networks, Alexandria, VA, pp. 59–70, 2006.

[19] D Liu, P Ning, Improving key predistribution with deployment knowledge in static sensor networks, ACM Transactions on Sensor Networks, V. 1, i. 2 (Nov 2005), pp. 204–239, 2005.

[20] R Luo, C Yih, K Su, Multisensor fusion and integration: Approaches, applications, and future research directions. IEEE Sensors J. 2, 2 (Apr 2002) 107–119.

[21] Mainwaring, J Polastre, R Szewczyk, D Culler, Wireless sensor networks for habitat monitoring, Proceedings of the 1st ACM International Workshop on Wireless Sensor Networks and Applications, Atlanta, GA, Sep 2002, pp. 88-97.

[22] K Malasri, L Wang, Addressing security in medical sensor networks, International Conference On Mobile Systems, Applications And Services, Proceedings of the 1st ACM SIGMOBILE international workshop on Systems and networking support for healthcare and assisted living environments, San Juan, Puerto Rico, pp. 7–12, 2007.

[23] K Martin, M Paterson, An Application-Oriented Framework for Wireless Sensor Network Key Establishment, Electronic Notes in Theoretical Computer Science, V. 192, i. 2 (May 2008), pp. 31–41.

[24] W Miller, (1989). Real-time application of neural networks for sensor-based control of robots with vision. IEEE Transactions on Systems, Man, and Cybernetics, 19(4): 825–831.

[25] L Oliveira, A Ferreira, M Vilaça, H Wong, M Bern, R Dahab, A Loureiro, SecLEACH-On the security of clustered sensor networks, Signal Processing, V. 87, i. 12 (Dec 2007), pp. 2882–2895.

[26] B Panja, S Madria, B Bhargava, A role-based access in a hierarchical sensor network architecture to provide multilevel security, Computer Communications, V. 31, I. 4 (Mar 2008),pp. 793–806.

[27] L Paradis, Q Han, A Survey of Fault Management in Wireless Sensor Networks, Journal of Network and Systems Management, V. 15, i. 2 (Jun 2007), pp. 171-190.

[28] J Riudavets, K Navarro, E Lawrence, R Steele, M Messina, Multi router traffic grapher (MRTG) for body area network (BAN) surveillance, Proceedings of the 4th WSEAS International Conference on Applied Informatics and Communications, Tenerife, Canary Islands, Spain, Article 71, 2004.

[29] R Shah, L Nachman, Interference Detection and Mitigation in IEEE 802.15.4 Networks, Proceedings of the 2008 International Conference on Information Processing in Sensor Networks, V. 00, pp. 553–554.

[30] L Schwiebert, S Gupta, J Weinmann, Research challenges in wireless networks of biomedical sensors, Proceedings of the 7th annual international conference on Mobile computing and networking, pp. 151–165, Jul 2001, Rome, Italy.

[31] C Strydis, D Zhu, G Gaydadjiev, Profiling of symmetric-encryption algorithms for a novel biomedical-implant architecture, Proceedings of the 2008 conference on Computing frontiers, Ischia, Italy, pp. 231–240, 2008.

[32] R Sugihara, R Gupta, Programming models for sensor networks: A survey, ACM Transactions on Sensor Networks, V. 4, i. 2 (Mar 2008), Article 8.

[33] D Takahashi, Y Xiao, F Hu, J Chen, and Y Sun, Temperature-aware routing for telemedicine applications in embedded biomedical sensor networks, EURASIP Journal on Wireless Communications and Networking, v. 2008, Article ID 572636, 11 pages.

[34] C Tan, H Wang, S Zhong, Q Li, Body sensor network security: an identity-based cryptography approach, Proceedings of the first ACM conference on Wireless network security, Alexandria, VA, RFID and embedded sensors, pp. 148–153, 2008.

[35] US FDA. Adverse event report to the US Food and Drug Administration, Center for Devices and Radiological Health Web site. http://www.accessdata.fda.gov/scripts/cdrh/cfdocs/cfMAUDE/Detail.CFM?MDRFOI__ID=465394. Accessed Jul 2008.

[36] R van der Togt, E van Lieshout, R Hensbroek, E Beinat, J Binnekade, P Bakker, Electromagnetic Interference From Radio Frequency Identification Inducing Potentially Hazardous Incidents in Critical Care Medical Equipment, JAMA, 2008; 299(24): 2884–2890.

[37] M Wang, M Blount, J Davis, A Misra, D Sow, A time-and-value centric provenance model and architecture for medical event streams, International Conference On Mobile Systems, Applications And Services, Proceedings of the 1st ACM SIGMOBILE international workshop on Systems and networking support for healthcare and assisted living environments, San Juan, Puerto Rico, pp. 95-100, 2007 .

[38] Ward, J.A., Lukowicz, P., Tröster, G. and Starner, T., Activity recognition of assembly tasks using body-worn microphones and accelerometers. IEEE Transactions on Pattern Analysis and Maching Intelligence. v. 28 i. 10. 1553–1567, 11 Aug. 2006.

[39] T Wark, C Crossman, W Hu, Y Guo, P Valencia, P Sikka, P Corke, C Lee, J Henshall, K Prayaga, J O'Grady, M Reed, A Fisher, The design and evaluation of a mobile sensor/ actuator network for autonomous animal control, Proceedings of the 6th international conference on Information processing in sensor networks, Cambridge, Massachusetts, pp. 206–215, 2007.

[40] H Zheng, ND Black and ND Harris(2005a), Position-sensing technologies for movement analysis in stroke rehabilitation, Medical & Biological Engineering & Computing, 43, 4, pp. 413–420.

[41] T Zia, A Zomaya, N Ababneh, Evaluation of Overheads in Security Mechanisms in Wireless Sensor Networks, Proc. of the 2007 International Conference on Sensor Technologies and Applications, pp. 181–185, 2007.

[42] Introduction to the Health Care Industry, Plunkett Research, Ltd., Retrieved from http://www.plunkettresearch.com/Industries/HealthCare/HealthCareTrends/tabid/294/ Default.aspx on 12/05/2008.

11

Security and Key Establishment Schemes for Distributed Sensor Networks

Than Dai Tran and Johnson I Agbinya

University of Technology, Sydney, Australia

In this chapter we study the security and key establishment schemes for distributed sensor networks (DSNs). Our approach is that first we provide an overview of all the research directions that constitute the security in DSNs. These include attacks, cryptography, key management, secure routing, secure applications, and intrusion detection. Then we will discuss in more detail about how the key agreement problem has been tackled via the investigation of key establishment and pre-distribution schemes. Finally, we will propose a new pairwise key establishment scheme for clustered DSNs. This scheme is shown to have desired security properties and be highly efficient, cost effective, and practical for resource-constrained sensor nodes.

11.1 Introduction

A distributed sensor network consists of thousands, even millions of tiny devices equipped with sensing, data processing, communication, and power components. These devices or sensor nodes also have additional application-specific units such as a location finding system, power generation, and mobilizer. They are often randomly and densely scattered over the sensing terrain cooperating with each other to complete pre-defined tasks. These nodes are assumed to be static after deployment and controlled by one or several control nodes, which collect data, monitor the status of and issue commands to sensor nodes [1].

DSNs have attracted lots of research interest due to their extensive and useful applications in military as well as civilian missions. These include, but are not limited to, detecting and tracking the movement of troops and military vehicles on a battlefield, law enforcement applications, automotive telemetric applications, room occupation monitoring in office buildings, measuring temperature and pressure in oil pipelines, and forest fire detection, monitoring environmental pollutants, measuring traffic flows on roads, etc. All these applications have unlimited benefits and potential; however, if security problems in such networks are not addressed properly and comprehensively, these benefits and potential are very likely to turn out to be bad risks to users.

Security challenges

Providing security in DSNs is a very challenging task due to the fact that they have many characteristics that make them very susceptible to attackers in adversarial environments such as a military battlefield. These challenging characteristics are summarised as follows:

- *Constrained resources*: Sensor nodes usually have severely limited processing capability, storage capacity, communication bandwidth, and power supply. As a result, the well-established security protocols developed for wired and ad hoc networks can not be ported directly to DSNs due to their complexity. Innovative security solutions taking the resource constraints into account are very necessary. The difficulty lies in the fact that a stronger security protocol costs more resources on sensor nodes, which can result in the performance degradation of applications. In contrast, weak security protocols can be broken easily by attackers. In most cases, a trade-off must be made between security and performance.

- *Hostile environments*: DSNs are often employed to execute tasks in hostile environments such as battlefields and operate in unattended manner. Therefore, sensor nodes can hardly avoid physical attacks. An attacker can capture sensor nodes; easily tamper with the captured nodes' memory to extract all their secrets that are used in security protocols since most nodes are not equipped with the tamper-proof feature due to the high implementation cost. He can also inject his own malicious nodes into the network to disrupt network operations.

- *Wireless communication*: Communication among sensor nodes is made possible via wireless channel. This channel is open in the sense that anyone can monitor or participate in providing he/she has a wireless interface operating at the same frequency band. This eases the way for attackers to break into the networks.
- *Lack of post-deployment network topology*: In most DSN applications, a sensor node does not know in advance which nodes will be in its communication range after deployment. This fact makes the task of designing and implementing security protocols more complicated.
- *Data fusion*: It is well-known that communication is the most costly operation in DSNs. On the other hand, due to the node density, vast amounts of duplicate sensed data exist in the network. Therefore, localised processing and data aggregation are needed to minimise the communication cost. The most suitable security method supporting this operation is to use a group key shared among the nodes in an immediate neighbourhood. However, in the context where nodes can be compromised, the confidentiality offered by the shared symmetric keys is not easy to maintain.

Security requirements

The paramount objective of security protocols in DSNs is to shield information, resources, and processes from threats and misbehaviour in the presence of resourceful attackers. Typically, accomplishing this objective requires one or more of the following security properties:

- *Availability*: DSNs are very vulnerable to the denial of service (DoS) attack such as radio jamming, network protocol disruption, or energy depletion. The availability implies that a DSN must be capable of providing services whenever they are required even in the presence of DoS attacks.
- *Authorisation*: This property ensures that appropriate rights are given to relevant nodes to join network operations.

- *Authentication*: This property ensures one communicating node can identify the source of information it receives from. Without authentication, an adversary can easily masquerade as a legitimate node to spread false information into the networks.
- *Confidentiality*: This property is used to provide the secrecy of important data transmitted over the air. That means nobody other than the intended receivers can have access to the content of a given message.
- *Integrity*: This property allows receivers to detect whether a received message is modified by malicious intermediate nodes or not.
- *Non-repudiation*: This property indicates that a node cannot deny sending a message it has previously sent.
- *Freshness*: This property assures that the received data is newly and ensures that no adversary can replay sent messages.
- *Resilience against physical attacks*: This property allows the network to be capable of tolerating the physical attacks such as node compromise, node replication to a great extent.

Furthermore, due to the fact that DSNs are dynamic in the sense that they allow addition and deletion of sensor nodes after deployment to extend the network or replace failing and unreliable nodes, it is suggested that forward and backward confidentiality should also be considered:

- *Forward confidentiality*: Sensor nodes are not capable of understanding any future messages after they leave the network.
- *Backward confidentiality*: Joining sensor nodes should not be able to access any past transmitted messages.

11.2   Distributed Sensor Network Security

Security for DSNs is an extensive research domain. It can be subdivided into a few categories including attacks, cryptography, key management, secure routing, secure applications, and intrusion detection. In this section, we discuss these sub-domains in more detail in order to build up an entire and concise picture of security in DSNs.

Attacks on DSNs

In broad view, attacks on DSNs can be classified into the following categories:

- *Outsider versus insider attacks*: Outsider attacks are defined as attacks mounted by nodes which are not deployed and managed by a legitimate authority. The impact of these attacks is limited. Insider attacks occur when an attacker can obtain authorisation to access the network. In this case, the insider attackers can cause more severe damage under the role of a legitimate entity. Typically, an attacker can become an insider by compromising a legitimate node or by deploying malicious nodes that are able to pass the network access control mechanisms.

- *Passive versus active attacks*: The goal of the passive attacks is to obtain information without being detected. An attacker often keeps silent to eavesdrop on the network traffic. Via passively participating in the network, the attacker can collect a large volume of traffic data and conduct analysis on the data to extract some secret information. As the way they work, it is very difficult to detect the passive attacks since the attacker does not leave much evidence. On the contrary, the active attacks exploits security flaws in the network protocol stack to launch a wide variety of specific attacks such as traffic jamming; packet modification, injection, replaying, or dropping. The active attacks induce more detrimental effects on the network than the passive attacks. Fortunately, these attacks are easier to detect since the attacker is actively involved in network communications and exposes much evidence caused by abnormal activities.

- *Mote-level versus laptop-level attacks*: The mote-level attacks refer to the ones launched using a few malicious nodes with similar capabilities to the legitimate nodes. Thus, the attacker has no more advantage in terms of resource supply. In the laptop-level attacks, the attacker can do much more harm to the network than a malicious sensor node, since it has much better power supply, as well as larger computation and communication capabilities than a legitimate node [2].

Table 11.1 Sensor network layers' vulnerabilities and denial-of-service defenses.

Network layer	Attacks	Defenses
Physical	Jamming	Spread-spectrum, priority messages, lower duty cycle, region mapping, mode change
	Tampering	Tamper-proofing, hiding
Link	Collision	Error-correctin code
	Exhaustion	Rate limitation
	Unfairness	Small frames
Network and routing	Neglect and greed	Redundancy, probing
	Homing	Encryption
	Misdirection	Egress filtering, authorisation, monitoring
	Black holes	Authorisation, monitoring, redundancy
Transport	Flooding	Client puzzles
	Desynchronization	Authentication

Approaching from the other angle, the attacks on DSNs can also be categorised as denial of service (DoS) attacks, attacks on routing, or node compromise based attacks. In [3], the authors classified various DoS attacks according to network layers. These attacks and possible defense techniques are summarised below:

Routing is an essential operation in DSNs. A number of routing protocols have been developed for DSNs. However, these protocols focused very much on efficiency and effectiveness of data dissemination and neglected to consider security issues in the design stage. Consequently, they are susceptible to attacks from the literature on routing in ad-hoc networks. As pointed out in [2], most network layer attacks against DSNs fall into one of the following categories:

- *Spoofed, altered, or replayed routing information*: An attacker may spoof, alter, or replay routing information in order to disrupt traffic in the network. These disruptions can create routing loops, attract or repel network traffic, extend or shorten source routes, generate fake error messages, partition the network, increase end-to-end latency, etc.
- *Selective forwarding*: Routing in DSNs is multi-hop and works based on the assumption that all nodes in the network accurately forward received messages. However, in hostile environments, malicious nodes may selectively forward only certain messages and simply drop others. An instance of this attack is the black hole attack in which a node drops all messages it receives.

- *Sinkhole*: This attack typically comes into play by making a compromised node look highly attractive to surrounding nodes via routing information manipulation. The resulting effect is that the surrounding nodes will select the compromised node as the forwarding node to route their data through.
- *Sybil*: The Sybil attack occurs when a malicious node deliberately presents more than one identity to the network. This attack has significantly negative impacts on fault-tolerant schemes, distributed storage, multipath routing, and network topology maintainance.
- *Wormholes*: In the Wormhole attack, an attacker can tunnel packets through a secret, low-latency link between two distant places and replay them. This attack can distort the network topology by making two distant nodes believe they are neighbors, thus it becomes a serious attack on routing protocols.
- *Hello flood attack*: Normally, HELLO packets are used in neighbour discovery phase. This means if a node receives the hello packet, it assumes that the sender is its immediate neighbour. A laptop-level attacker may use high-powered transmitter to convince a large area of nodes that they are in the neighbourhood of the transmitting node.
- *Acknowledgment spoofing*: Some routing protocols for DSNs rely on link layer acknowledgments. A malicious node can spoof the acknowledgments of overheard packets addressed to neighbouring nodes. Objectives include convincing the sender that a weak link is strong or that a dead or disabled node is alive.

Hostile environments and unattended conditions expose sensor nodes to the node compromise attack. As the result, an attacker can further mount several malicious attacks such as node replication [4, 5], node collusion [6, 7], key misuse [8], etc.

Cryptography in DSNs

Like its counterpart in the traditional networks, cryptography in DSNs falls into two trends: public key and symmetric key. Due to the resource constraints, cryptographic methods for DSNs should be designed deliberately and be evaluated by code size, data size, processing time, and energy consumption.

Many pieces of research work have revealed that traditional public key algorithms such as RSA are prohibitively costly to be employed in DSNs in comparison to symmetric key ones [9–11]. As demonstrated in [9], the encryption of a 1024-bit block consumes approximately 42 mJ on the MC68328 DragonBall processor using RSA, while the estimated energy consumption for a 128-bit AES block is a much lower at 0.104 mJ. Despite of this fact, recent efforts have been made to show that it is feasible to apply public key cryptography to DSNs by using the proper selection of algorithms and associated parameters, optimisation, and low-power techniques related to both hardware and software aspects [12–20]. The investigated public key algorithms include Rabin's scheme, NtruEncrypt, RSA, and Elliptic Curve Cryptography (ECC). Most of the studies in literature focus on RSA and ECC algorithms. The advantage of ECC is that it appears to offer equal security for a much smaller key size, thereby decreasing computational and communication overheads.

However, because the constraints on computation, power supply, and communication bandwidth in sensor nodes limit the application of public key cryptography, most research studies have concentrated on symmetric key one in DSNs. In [21], five well-known encryption schemes, RC4, RC5, IDEA, SHA-1, and MD5 were measured on six different architectures ranging in word size from minimum 8 bit to maximum 32 bit width to determine the impact of embedded architectures on their performance. This will help designers predict the performance of a system for cryptographic tasks. In [11], Law *et al.* conducted a survey based on existing literature and authoritative recommendations to identify the candidates of block ciphers suitable for sensor networks. They are Skipjack, RC5, RC6, Rijndael, Twofish, MISTY1, KASUMI, Camellia. These candidates are evaluated and accessed based on not only the security properties but also the storage- and energy-efficiency. Finally, the authors discovered that the most suitable ciphers for sensor networks are Skipjack, MISTY1, and Rijndael, depending on the combination of available memory and required security. In terms of operation mode, the Output Feedback Mode is recommended for pairwise links and Cipher Block Chaining for group communications.

Key management protocols

Key management is a fundamental mechanism to ensure the security of network services and applications in DSNs. The aim of key management is to establish and secure required keys between communicating sensor nodes. Typically, in order to achieve this aim, four processes are involved including key setup, key distribution/redistribution, key generation, and key revocation.

Due to the fact that most communication patterns in DSNs are node-to-node communication, most key management protocols boil down to key establishment ones. Traditionally speaking, key establishment is a process or protocol whereby a shared secret becomes available to two or more parties, for subsequent cryptographic use. The proposed key establishment mechanisms for wireless sensor networks differ in various aspects, which are not surprising because DSNs operate in vastly different environments with different levels of security threat, supporting different applications with different communication patterns and network topologies and using different sensor nodes with different resource capabilities. These key establishment schemes can be classified broadly according to employed cryptographic techniques including symmetric and asymmetric cryptography. The schemes can also be categorized based on methods whereby shared keys are generated and delivered into key transport and key agreement protocols. In addition, they can be classified into static or dynamic approaches based on whether rekeying (update) of administrative keys is supported post network deployment [22]. Another classification criterion can be used is the role of individual network entities in the key management process. In a homogeneous scheme, all nodes perform the same functionality; on the other hand, nodes in a heterogeneous scheme are assumed different roles. Homogeneous schemes are generally designed for flat network architecture; meanwhile heterogeneous schemes are intended for both flat and clustered network topologies. Note that in the taxonomy shown in Figure 11.1, only shinning representatives of each type of the key establishment schemes are discussed and taken into account.

Symmetric Techniques vs. Asymmetric Techniques

Like its traditional counterpart, the key establishment problem in wireless sensor networks can be approached through symmetric-key techniques or public-key techniques.

Symmetric techniques: In this approach, the key establishment is performed via the aid of symmetric-key primitives such as symmetric-key ciphers, arbitrary length hash functions (MACs), signatures, and pseudorandom sequences. In addition, symmetric-technique based key establishment

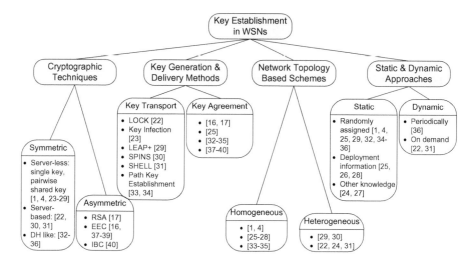

Fig. 11.1 Classification of key establishment schemes in WSNs.

schemes can also be identified according to key (keying) distribution models such as server-less schemes including single network-wide key scheme, pairwise shared keys scheme [1, 4, 23–29]; server-based schemes [22, 30, 31]; and Diffie-Hellman like schemes [32–36].

Asymmetric techniques: In contrast to the proliferation of symmetric-key based methods, applications of asymmetric (public-key) techniques in resolving the key establishment problem for WSNs have remained in the initial stage due to the limited capabilities of sensor nodes in terms of memory storage, computation, communication and power supply. Most of the research efforts focus on demonstrating feasibility of public-key algorithms on sensor platforms via requirements of special hardware support, algorithm and instruction-level optimization for particular architectures [14, 17, 37–39]. Public-key based key establishment schemes can be subdivided into several types based on the utilized cryptography such as RSA [17], elliptic curve cryptography (ECC) [16, 37–39], identity-based cryptosystem (IBC) [40].

One straightforward observation drawn from this classification is that symmetric-key approaches for establishing shared keys in WSNs are still preferred choices since they involve only simple cryptographic operations, haves low cost, and can be implemented independently of the architecture.

Key Transport Protocols vs. Key Agreement Protocols

Key transport protocols: Classically speaking, a key transport protocol or mechanism is a key establishment technique where one party creates or otherwise obtains a secret value, and securely transfers it to the other(s) [41]. In this sense, there are a number of schemes and protocols for WSNs worth being named such as LOCK [22], key infection [23], LEAP+ [29], SPINS [30], SHELL [31], and path key establishment [33, 34].

Key agreement protocols: As for key agreement schemes, during an initialization stage, secret data values (keying materials) are generated by a trusted server and distributed to users, such that any pair of users may subsequently compute a shared key unknown to all others (aside from the server). By this definition, the schemes in [16, 17, 25, 32–35, 37–40] can be taken into account.

Static Schemes vs. Dynamic Schemes

Static schemes: In these schemes, administrative and link keys are assumed to be preloaded into the nodes once for their whole life, they will not be updated by the time. Administrative and link keys are generated prior to deployment, assigned to nodes randomly [1, 4, 25, 29, 32, 34–36], based on some deployment information [25, 26, 28], or other knowledge [24, 27] and then distributed to nodes. When these keys are disclosed due to node compromise, they are revoked and deleted from related sensor nodes via key or node revocation schemes [1, 4, 42].

Dynamic schemes: Dynamic key management schemes may change administrative keys periodically [36], on demand or on detection of node capture [22, 31]. The major advantage of dynamic keying is enhanced network survivability, since any captured key(s) is replaced in a timely manner in a process known as rekeying. Another advantage of dynamic keying is providing better support for network scalability; upon adding new nodes, unlike static keying, which uses a fixed pool of keys, the probability of network capture does not necessarily increase [22]. However, their downside is also worth considering. In order to operation properly, these schemes require either powerful hardware components which are sometimes expensive to satisfy or/and costly software functionalities such as time synchronisation and intrusion detection systems.

Network topology dependent schemes

Homogeneous schemes: As mention earlier, in homogeneous schemes, the role of each network node in establishing shared keys is identical except base stations. More specifically, each node is pre-distributed some keys or keying materials before sensor deployment. These keys or keying materials are used by adjacent communicating nodes on the fly after deployment to establish shared keys in absence of superior nodes [1, 4, 25–28, 33–35]. Such schemes are widely used applications that make use of the flat network topology wherein hop-by-hop communication is the dominant communication pattern in DSNs.

Heterogeneous schemes: Heterogeneous schemes are designed to support multi-tier architectures wherein different network nodes take on different network missions. For example, normal sensor nodes are used to sense interested events and deliver their readings to a cluster head (or data fusion node, aggregation point). The readings are then processed and aggregated there before being forwarded towards the base station. Since there are several types of communications between different network nodes of different roles, single keying mechanism is not suitable for meeting security requirements. Typically, heterogeneous schemes support the establishment of several types of keys: a pairwise key shared between two neighboring sensor nodes, a pairwise key shared by a sensor node and a cluster head, an individual key shared between a sensor node and the base station, a cluster key shared by all nodes in the cluster, and a global key shared by all the nodes in the network [29, 30]. Furthermore, cluster heads are further assumed to be more powerful than normal sensor nodes and have extra functionalities to ease the burden of the key management process [22, 24, 31].

Secure routing protocols

Routing protocols are the most important element in DSNs since they are designed to tackle the problem of constructing a route from a source to a destination and delivering data on the constructed route. DSN routing protocols are very application-specific. They can be classified into three themes according to network topologies: flat-based routing, hierarchical-based routing, and location-based routing [43]. However, the problem is that most routing protocols lack integrating security into their design. In harsh environments, this induces the routing protocols to highly suffer from security threats as discussed in 2.1. Therefore, securing routing is very crucial task to mitigate or avoid these threats.

Few possible security countermeasures against the routing threats have been discussed in the literature. Link layer encryption and authentication

[2, 30, 44–46] can be used to protect DSNs against eavesdropping, injection, modification attacks. To counter selective forwarding and node compromise attacks, redundant multi-path routing can be employed [2, 44]. Node impersonation attack can be prevented using authentication through a base station [30]. False routing information can be detected via broadcast authentication mechanisms [30, 47] and topology verification by the base station [44]. To counter DoS attacks, individual nodes are not permitted to broadcast information to the entire network [44].

Secure applications

In hostile environments, DSN applications are more likely to suffer from various security issues and thereby may not operate in a desired manner. In this section, these issues are discussed together with counteractive techniques.

The first security issue is false report injection. When an interested event occurs, surrounding sensor nodes have similar observations of the event. In order to mitigate heavy traffic and conserve energy caused by the transmission of the redundant sensed data back to the base station from all the nodes, the in-network processing is employed. Accordingly, the data redundancy is eliminated by and a final report is generated and sent by a data fusion node/data aggregator to the base station. However, if some nodes are compromised and then inject false data into the network, this data fusion mechanism will be disrupted. The counteractive techniques can be report verification [48], en-route filtering [49–51], hop-by-hop aggregation [52–54], and resilient fusion [55].

The second security issue is false location report. The knowledge of physical sensor locations is indispensable to many location-based applications such as target detection and tracking, precise navigation, search and rescue, geographic routing, security surveillance, and so on. However, the effectiveness of these applications is threatened by malicious nodes which can claim false location information. The countermeasures against this threat can be the echo protocol [56], a secure range-independent localisation scheme (SeRLoc) [57], the verifiable multilateration [58–60] and so forth.

The third security issue is the compromise of the source location privacy. The confidentiality of the source location in some scenarios is very crucial because otherwise attackers can find out the location of valuable monitored targets by tracing back to the source node. To provide the source location privacy, a new routing protocol, termed phantom routing, has been introduced in [61, 62]. This routing is alleged to increase the source location privacy while negligibly increasing the communication overhead compared with the flooding and the single-path routing.

Intrusion detection in DSNs

Despite the fact that multiple security mechanisms have been developed to secure DSNs, attackers can still be capable of infiltrating into the networks via node compromise and mounting the insider attacks due to the lack of physical security together with unattended operations. In this situation, preventive mechanisms such as encryption and authentication are not powerful enough to guard DSNs against malicious insiders. Therefore, intrusion detection systems (IDSs), serving as the second line of defense, are vital for providing a highly secured information system [63, 64].

An IDS is a security technology that attempts to detect and identify intruders who are trying to break into and misuse a system without authorisation and malicious users who have legitimate access to the system but are abusing their privileges. Typically, there are two approaches to the intrusion detection: rule-based detection and anomaly-based detection. The rule-based IDSs are used to detect intruders based on known patterns of intrusions and while anomaly-based systems detect new or unknown intrusions [63].

The research on intrusion detection in DSNs is still preliminary due to particular obstacles such as resource constraints, lack of information on topology, normal usage, traffic patterns, and so on. In [65], Silva et al. developed a decentralised IDS based on the interference of the network behaviour obtained from the analysis of events detected by a intrusion detection agent. These events can be data message listened to by the agent that is not addressed to it, and message collision when the agent tries to send a message. Onat and Miri [66] introduced an IDS where sensor nodes have the ability to record simple statistics about their neighbors' behavior and detect anomalies in them. Su *et al.* [67] proposed an intrusion detection approach to detect and revoke the compromised nodes in clustering-based sensor networks with energy-saving consideration. Martynov [68] designed and implemented an agent-based IDS for sensor networks that can detect potential DoS attacks using anomaly patterns. Cheng et al. [69] recently proposed an application-independent detection model, distributed cross-layer detection model (DCD), making use of a distributed mechanism and the information of each layer in the communication protocol to detect already compromised sensor nodes.

11.3 Key pre-distribution schemes for sensor networks

Generally key pre-distribution schemes are key establishment protocols whereby the resulting established keys are completely determined a priori by initial keying material [41].

In sensor networks, most of the key establishment proposals make use of the fact that WSNs virtually operate under administration of a single central authority to ease key deployment tasks. Accordingly, before deployment, sensor nodes are pre-distributed either keys directly or keying materials which are used after deployment to dynamically generate pairwise keys among communicating nodes. The key challenge is how to devise an efficient method for distributing keys and keying materials to sensor nodes prior to deployment such that an optimal compromise between conflicting requirements of optimal resource usage, network scalability, high shared-key connectivity and network resilience could be reached.

Approaches to key pre-distribution problem in WSNs can be classified broadly into three main categories: (i) random and variants, (ii) deterministic, or (iii) hybrid. In the first category, a key ring is randomly drawn out of a large key pool without replacement and distributed into the memory of each sensor node. In addition, several techniques and knowledge have been proposed to improve one or several performance metrics at the cost of the other metrics. In the second category, deterministic algorithms are utilized to generate keying materials for sensor nodes. Using these keying materials, communicating nodes can definitely derive a shared key between them. Lastly, the hybrid approach combines merits of the two first approaches to improve network resiliency while still maintaining shared-key connectivity at a very high level.

Mathematical Model

At the highest level of abstraction, a WSN is modeled as a set of N nodes $\Gamma = \{S_i | i = \overline{1, N}\}$ most of which have the same hardware configuration and software functionalities. Then a set $\Psi = \{\kappa_i | i = \overline{1, P}\}$ of keys or keying materials is generated following a specific key pre-distribution scheme. Each sensor node S_i is assigned a subset $\Lambda_i = \{\kappa_{ij} | i = \overline{1, N}, j = \overline{1, k}\}$ derived from the set Ψ prior to node deployment. This assignment is performed in such a way that the intersection set of any pair of subsets $\Lambda_i, \Lambda_j \subset \Psi$ is not empty. That is, $\Lambda_i \cap \Lambda_j \neq \emptyset$. After deployment, sensor nodes will use key or keying elements in $\Lambda_i \cap \Lambda_j$ to establish pairwise keys for securing communication links.

Probabilistic Key Sharing Scheme

This scheme was proposed by Eschenauer and Gligor [1] which is also referred to as *basic scheme*. It relies on probabilistic key sharing among the nodes of a random graph and uses a simple shared-key discovery protocol for key establishment. Before sensor network deployment, a key ring of k keys is selected randomly without replacement from a large pool of P keys which is generated off-line and distributed to each sensor node together with respective key identifiers. These key identifiers then might be used by any pair of neighboring nodes to examine whether they have a shared key during the *shared-key discovery* phase. As alternate method for shared key discovery, a sensor node may challenge its communicating partner to solve puzzles instead of disclosing the key IDs to secure key-sharing patterns.

Due to the randomness of key ring drawing, a shared key might not exist between some pairs of nodes. Fortunately, as long as a path of nodes sharing keys pair-wise exists between two nodes, the pair of nodes can use that path to exchange a key to establish a secure direct link during *path-key establishment* phase.

The feasibility of basic scheme is supported by random graph theory [70]. Accordingly, given a desired probability P_c for shared-key connectivity, network size n, and network density n' one can find out the proper values of k and P using the following formula derived from random graph theory:

$$\left(\frac{n-1}{n'-1}\right) \times \left(\frac{\ln(n) - \ln(\ln(-P_c))}{n}\right) = 1 - \frac{((P-k)!)^2}{(P-2k)!P!} \quad (11.1)$$

It is shown, for example, that to establish "almost certain" shared-key connectivity for a 10,000-node network, a key ring of only 250 keys drawn of have to be pre-distributed to every sensor nodes.

Advantages of this scheme include flexible, efficient, and fairly simple to implement, while also offering good scalability. Disadvantages of this scheme include vulnerability to small-scale node capture attack and lack of node-to-node authentication property.

Extended Versions of The Basic Scheme

Chan *et al.* [4] further extended the basic scheme into two mutually exclusive variants *q-composite keys scheme* and *multipath key enforcement*. The goals of these variants are both to strengthen the security of an established link key by either increasing the amount of key overlap of q common keys ($q > 1$) required for key setup or establishing the link key through multiple paths. In the former, the main problem is to decide the size of key pool. On one hand, it should be reduced to preserve the given probability p of two nodes sharing at least q keys to establish a secure link. On the other hand, it should be large enough to lower attackers' gain of a larger sample of key pool by breaking fewer nodes. The latter is a security supplement to the basic scheme. The aim is to update a pairwise key k established during initial key setup of the basic scheme between two neighboring nodes A and B. To realize this, multiple secure link-disjoint paths between A and B need to be discovered first. Let assume j be the number of such paths. A then generates j random values $n_1, \ldots, n_j$ of the same length and securely send them to B. Then the new link key can be computed by both A and B as:

$$k' = k \oplus n_1 \oplus n_2 \oplus \cdots \oplus n_j \qquad (11.2)$$

Random Pairwise Keys Scheme

One severe security problem of the basic scheme is the lack of node-to-node authentication. Because keys can be issued multiple times out of the key pool, many nodes may hold this set of secret keys K in their key ring. Hence, a node cannot ascertain that it is really communicating with one party or another party since it knows nothing more about its communicating party than its knowledge of K.

In an effort to enhance the basic scheme, Chan, *et al.* [4] proposed the random pairwise keys scheme with the key property of node-to-node authentication. The two phase scheme is summarised as follows. In the initialization phase, unique IDs are generated for a sensor network of up to $n = \frac{m}{p}$ nodes. Each node is assigned different m pairwise keys to m other randomly selected distinct nodes. The keys are stored in the node's key ring along with the corresponding node IDs. In the post-deployment key setup phase, each node first broadcasts its ID to its one-hop neighbours. The neighbouring nodes are able to determine a pairwise key shared with the broadcasting node if they find its ID in their key rings. According to this bootstrapping method, apparently, the probability of a pairwise key existing between two nodes is p. This probability is calculated such that the network connectivity is achieved with high probability using random graph theory [70].

Polynomial Pool-Based Key Predistribution

This framework for key predistribution was developed by Liu and Ning [34]. Basically, it can be considered as the combination of polynomial-based key predistribution, one kind of deterministic key predistribution and the key pool idea used in [1, 4].

The polynomial-based key predistribution is a special case of group key distribution protocol developed by Blundo *et al.* [71] in the context of sensor networks. In this scheme, to predistribute pairwise keys, the (key) setup server randomly generates a bivariate t-degree polynomial $f(x, y) = \sum_{i,j=0}^{t} a_{ij} x^i y^j$ over a finite field F_q where q is a prime number that is large enough to accommodate a cryptographic key, such that it has the property $f(x, y) = f(y, x)$. For each sensor i, the setup server computes a polynomial share of $f(x, y)$, that is, $f(i, y)$. For any two sensor nodes i and j, node i can compute the common key $f(i, j)$ by evaluating $f(i, y)$ at point j, and node j can compute the same key $f(j, i) = f(i, j)$ by evaluating $f(j, y)$ at point i. However, this scheme has a security weak point if applied directly to sensor networks. That is the collusion of more than t compromised sensor nodes results in the revelation of any pairwise key between any two non-compromised nodes.

The proposed framework has three general phases: setup, direct key establishment, and path key establishment. The setup phase is used by the setup server to pick a subset of polynomials and pre-distribute the shares of these polynomials to each sensor node. After being deployed, if two sensor nodes need to establish a pairwise key, they first attempt to do so via direct key establishment phase. In the case this phase does not work then the two nodes start path key establishment phase.

Based on the ways to perform the subset assignment problem, the authors presented three instantiations of the framework: random subset assignment key predistribution scheme, grid-based key predistribution scheme, and hypercube-based key predistribution scheme. In essence, the first instantiation works in the same manner as the basic scheme does if each key in a key ring in the basic scheme is substituted by a polynomial share in the random subset assignment scheme. Meanwhile, the subset assignment of the last two instantiations was developed based on the idea proposed by Gong and Wheeler [72].

Multiple-Space Key Pre-distribution Scheme

In [33], Du *et al.* developed multiple-space key pre-distribution scheme in a similar approach presented in the polynomial pool-based key predistribution to solve key distribution problem for WSNs. The scheme builds on Blom's key pre-distribution scheme [73] and combines the idea of key pool and key rings in the basic scheme with it.

Specifically, the scheme first uses Blom's scheme as a building block to generate multiple key spaces, a pool of tuple (D, G), where matrices D and G are as defined in Blom's scheme. Then this pool is used as a pool of keys as in the basic scheme to establish a common secret key between any pair of nodes.

Key Management Schemes Using Deployment Knowledge

Du *et al.* [26] proposed exploiting node deployment knowledge based on group-based deployment model together with the idea of key pool in the basic scheme to improve the performance of the previous random key pre-distribution schemes. Node deployment knowledge is defined as the knowledge about the nodes that are likely to be the neighbours of each sensor node. This knowledge can be modelled using non-uniform probability density functions (pdfs) such as normal Gaussian distribution.

Making good use of the knowledge, the authors proposed to divide the common key pool into different key pools. Each of them is corresponding to on deployment group. The goal of setting up the keys pools is to allow the nearby key pools which are corresponding to deployment groups of neighboring locations to share more keys, while pools far away from each other share fewer keys or no keys at all. By doing like this, each node only needs to carry a fraction of the keys required by the other key pre-distribution scheme [1, 4] while attaining the same level of shared-key connectivity. Furthermore, this reduction in key storage substantially improves network's resilience against node capture.

Location Dependent Key Management Scheme

F. Anjum [27] proposed a location aware approach for key management in sensor networks. This approach, location dependent key (LDK) management, is aimed at improving network resiliency against node compromise by diversifying network-wide pre-distributed keys on each key ring of a sensor node into location-specific keys.

This scheme has two noticeable advantages. First, it confines the impact of node compromise to the vicinity of compromised nodes only. This means that compromise of a node in a location affects the communications only around that location. Furthermore, this confinement also helps the scheme avoid two malicious attacks: the node replication attack [5, 74, 75] and the key-swapping collusion attack [6, 7]. Second, the scheme guarantees that sensor nodes in different locations have different keys without resort to any knowledge about the deployment of sensor nodes.

However, several significant disadvantages make the scheme impractical. Firstly, the assumption of the presence of anchors is not always reasonable. Moreover, the requirement of the tamper proof feature from anchors adds more cost to the scheme. Secondly, in order to support incremental node addition, either the anchors are required to either transmit beacons periodically or newly deployed sensor nodes have to send signals to the anchors. These are very problematic requirements since the former results in quick energy depletion of the anchors and the latter facilitates the energy depletion attack on the anchors. Thirdly, the transmission pattern of the anchors make easier for attackers to locate the anchors for compromise.

Random Perturbation-Based Scheme

W. Zhang *et al.* [35] developed a random perturbation-based scheme for pairwise key establishment in sensor networks in an effort to address several limitations of existing pairwise key establishment schemes including no guarantee for direct key establishment, no resiliency to large-scale node compromise, no tolerance to dynamic network topology.

In fact, this scheme is a variant of the basic polynomial-based scheme [71] in which the scheme does not give each node an original share of a symmetric polynomial but the perturbed share, which is the sum of the original share and a perturbation polynomial introduced by the authors themselves. The benefit of this addition is twofold. On one hand, two nodes can still establish a key using the perturbed shares. On the other hand, the adding introduces prohibitively high complexity to an attacker in order to break the symmetric polynomial even if he/she has compromised a large number of sensor nodes. This means that any number of compromised colluding nodes have negligible probability to break the pairwise key established by a pair of non-compromised nodes.

Despite the above noteworthy features, this scheme is still highly vulnerable to the node replication attack and the key-swapping collusion attack.

11.4 Non-threshold Deterministic Parwise Key Establishment Scheme for Clustered Distributed Sensor Networks

In this section, we design a pairwise key establishment (PKE) for clustered DSNs. The motivation is based on the fact that although random key pre-distribution schemes (RKPSs) [1, 4, 24] and their improved variants threshold RKPSs [32–34] have been considered as one of the most practical PKE approaches to tackling key agreement problem in DSNs, they still have a few security flaws. First, the RKPSs are venerable to even a small-scale node compromise [4] whereas regardless of security improvements the threshold PKPSs still possess an undesirable threshold property that allows an attacker after having obtained a certain number of collusive compromised sensor nodes to compromise any pairwise key established between uncompromised communicating nodes. Second, they do not succeed in guaranteeing that any two nodes can establish a pairwise key at any time especially in the presence of node failure. Lastly, in many cases, they may still resort to the path key establishment technique which requires revealing secrets to or obtaining secrets from any third parties. The involvement of the third parties is highly undesirable since they may be compromised and more resource consumption is unavoidable due to message encryption, decryption, and transmission.

The proposed scheme exploits the clustering topology of DSNs to overcome the aforementioned limitations. It exhibits the following features:

1. Any two communicating nodes can always establish a pairwise key whenever needed without need of third parties' involvement and irrespective of the network size, node density, deployment knowledge, and node mobility.
2. Our scheme is immune from large number of node compromise, threshold collusion attack, and fake ID attack by nature.
3. Through analysis, our scheme is demonstrated to be computationally efficient, require low storage cost, and posses little communication overhead.

The rest of this chapter is arranged as follows. First the network and threat models are introduced followed by the background from which our scheme is developed. Then our scheme will be described. Finally, the performance and security properties will be analyzed.

Network and Threat Models

Network model

In our scheme, we consider a distributed sensor network comprising a base station and enormous sensor nodes which are organized into clusters as illustrated in Figure 11.2. Each cluster is governed by a cluster head, which can broadcast messages to all sensor nodes in the cluster. Each cluster head is assumed to be reachable to all sensors in its cluster, either directly or in multi-hops. The clusters of sensors are deployed at deployment points using airplanes as mentioned in, helicopters or unmanned vehicles. All the clusters of sensors could be deployed at a time or more clusters could be deployed later on the basis of demand.

Each kind of network entities possesses different capabilities. The base station is assumed to be very resource-rich and fairly far from the sensor field.

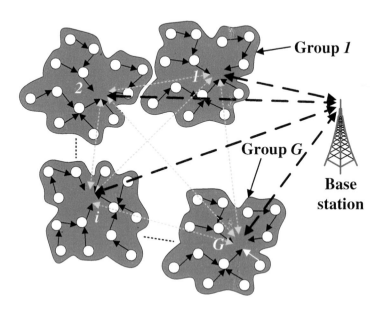

Fig. 11.2 Clustered distribution sensor networks model.

The cluster heads are relatively powerful with sufficient energy and communication range to transmit data to the base station and the other cluster heads. Finally, the sensor nodes are low-cost, compact, lightweight, disposable, and constrained in energy, computation, and communication resources. Nodes in different clusters can communicate with nodes in other cluster through cluster heads.

Threat Model

In this paper, we mainly consider sensor networks being deployed in hostile areas where an adversary tries to manipulate the system through capturing and compromising some sensor nodes. When sensor nodes are captured, their memory can be read, erased or tampered with. Therefore, the adversary would know keying materials of compromised nodes. Using those materials, the adversary could mount all kinds of attacks mentioned in the literature such as replication attack [4, 5, 74], collusion attack [6, 76], misused keys related attacks [8] and other routing related attacks [2]. However, the two last types of attacks are out of the scope of this paper and considered being solved by proposals in the existing literature [2, 8]. In addition, cluster heads are supposed not to be compromised easily. Since the number of cluster heads is quite few, their compromise is further assumed to be detected and identified promptly.

Background: Matsumoto-Imai Scheme

This subsection explains how the network nodes' secret keying materials are generated and how network nodes use these materials to establish pairwise keys in the manner of the Matsumoto-Imai scheme [76].

A central server first generates $l(m \times m)$ symmetric matrices M_ωs over finite field $GF(2)$. M_ω is used to generate the ω-th bit of a pairwise key between any pair of neighboring nodes, so l is the length of this key. The central server then computes the keying material Φ_i for each node S_i as follows:

$$\Phi_i^\omega = y_i M_\omega (\omega = \overline{1, l}) \tag{11.3}$$

$$\Phi_i = \begin{bmatrix} \Phi_i^1 & \Phi_i^2 & \cdots & \Phi_i^l \end{bmatrix}^T \tag{11.4}$$

where $y_i (i = \overline{1, N})$ is the m-dimensional vector, effective ID of node S_i. Φ_i is the private information and kept secret from both other sensor nodes and attackers. Φ_i^ω and Φ_i are illustrated in the Figure 11.3.

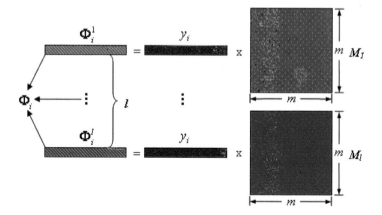

Fig. 11.3 Structure of sensor node Si's keying material Φ_i.

By using Φ_i and Φ_j, two neighboring nodes S_i and S_j compute their pairwise key as follows:

$$S_i \quad \begin{aligned} K_{ij}^\omega &= \Phi_i^\omega y_j^T \ (\omega = \overline{1,l}) \\ K_{ij}^T &= \Phi_i y_j^T \end{aligned} \quad (11.5) \qquad S_j \quad \begin{aligned} K_{ji}^\omega &= \Phi_j^\omega y_i^T \ (\omega = \overline{1,l}) \\ K_{ji}^T &= \Phi_j y_i^T \end{aligned} \quad (11.12)$$

where symbol T denotes transposition operation.

For the sake of clarity, we give a simple example to illustrate how two nodes Si and Sj compute the first bit of a pairwise key given that:

$$\omega = 1, \ M_1 = \begin{bmatrix} 1 & 0 & 0 \\ 0 & 1 & 1 \\ 0 & 1 & 0 \end{bmatrix} y_1 = \begin{bmatrix} 0 & 0 & 1 \end{bmatrix}, \ y_2 = \begin{bmatrix} 0 & 1 & 0 \end{bmatrix}$$

The computation is as follows:

$$\Phi_1^1 = y_1 M_1 = \begin{bmatrix} 0 & 0 & 1 \end{bmatrix} \begin{bmatrix} 1 & 0 & 0 \\ 0 & 1 & 1 \\ 0 & 1 & 0 \end{bmatrix} = \begin{bmatrix} 0 & 1 & 0 \end{bmatrix}$$

$$K_{12}^1 = \Phi_1^1 y_2^T = \begin{bmatrix} 0 & 1 & 0 \end{bmatrix} \begin{bmatrix} 0 \\ 1 \\ 0 \end{bmatrix} = 1 \Rightarrow \boxed{K_{12}^1 = 1}$$

$$\Phi_2^1 = y_2 M_1 = \begin{bmatrix} 0 & 1 & 0 \end{bmatrix} \begin{bmatrix} 1 & 0 & 0 \\ 0 & 1 & 1 \\ 0 & 1 & 0 \end{bmatrix} = \begin{bmatrix} 0 & 1 & 1 \end{bmatrix}$$

$$K_{21}^1 = \Phi_2^1 y_1^T = \begin{bmatrix} 0 & 1 & 1 \end{bmatrix} \begin{bmatrix} 0 \\ 0 \\ 1 \end{bmatrix} = 1 \Rightarrow \boxed{K_{12}^1 = 1}$$

The Proposed Scheme Description

Cluster and Node Identification Assignment

Continuing to develop the assumption in the network model section, we assume the entire network of N nodes is grouped into G clusters where G is the maximum number of deployable clusters. Each cluster C_i has N_i nodes. That is:

$$\sum_{i=1}^{G} N_i = N \tag{11.7}$$

The identification ranges of nodes including cluster heads in different clusters are assigned in such a way that they are not overlapped. Specifically, the unique ID of each ith node of the jth cluster S_{ij} comprises a cluster ID segment, an intra-cluster node ID segment, and zero-padding segment as illustrated in the Figure 11.4.

As shown in the figure, the identification is represented by m_j bit effective ID in the node's memory. The zero-padding segment is optional and only added to guarantee that $m_j > \max\{N_i, i = \overline{1, G}\}$.

Keying material pre-distribution

Base station: Base station is pre-distributed two types of key corresponding to two types of communications between base station and cluster heads. These

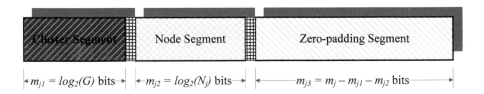

Cluster Segment	Node Segment	Zero-padding Segment
$\leftarrow m_{j1} = log_2(G)$ bits $\rightarrow$	$\leftarrow m_{j2} = log_2(N_j)$ bits $\rightarrow$	$\leftarrow m_{j3} = m_j - m_{j1} - m_{j2}$ bits $\rightarrow$

Fig. 11.4 Node identification structure.

are pairwise keys and a level one broadcast key. Each of the pairwise keys is used to secure communications between the corresponding cluster head and base station. The level one broadcast key is a group key used to secure communications among base station and all cluster heads. Therefore, base station stores totally $G + 1$ keys.

Cluster heads: Each cluster head is pre-loaded with two keys: a pairwise key shared with base station only and the level one broadcast key shared with not only base station but also the other cluster heads. Moreover, each cluster head is also pre-stored a level two broadcast key for securing broadcast messages within the cluster and keying materials used to establish pairwise keys with sensor nodes in its cluster and other cluster heads. Details of the keying material generation are described below.

Sensor nodes: Each sensor node is pre-distributed the following information:

- The level two broadcast key as mentioned above.
- Keying material for pairwise key establishment between itself and another entity in its cluster including its cluster head.
- Table 11.2 summarizes the cryptographic information required by communication tasks of each network entity type.

Keying material generation: The keying material generation must assure that two things have to be done: successful intra-cluster pairwise key establishment between any pair of neighboring sensor nodes within each cluster and successful inter-cluster pairwise key establishment between any pair of cluster heads. Therefore, the following business has to be taken into account:

Table 11.2 Network entities, communication tasks and required cryptographic information.

Network entity	Communication tasks	Cryptographic information
Base station	Broadcast to all cluster heads	Level one broadcast key
	Unicast to cluster heads	Pairwise keys
Cluster head	Broadcast to base station and other cluster heads	Level one broadcast key
	Broadcast within cluster	Level two broadcast key
	Unicast to base station	Unicast to base station
	Unicast to another cluster head	Keying material (type 1)
	Unicast to a sensor node	Keying material (type 2)
Sensor node	Broadcast within cluster	Level two broadcast key
	Node-to-node communication	Keying material (type 2)

1. For each cluster C_h, the central server has to generate $l(m_h \times m_h)$ symmetric matrices and compute the keying material Φ_i (renamed as Φ_{ih}) for each node S_{ih} (including cluster head) based on its effective ID as described in the background section. The l symmetric matrices used for one cluster are different from those for other clusters. The value of m_h is determined such that $m_h > N_h$. The explanation of this inequality constraint is provided later.

2. For secure inter-communication among clusters, the central server also generates $l(m_c \times m_c)$ symmetric matrices and compute the keying material Φ_i (renamed as Φ_{ic}) for each cluster head S_{0i}. These l symmetric matrices are different from the ones mentioned above. The value of m_c is determined such that $m_c > G$. The reason of this inequality constraint is given later.

Therefore, each cluster head contains two kinds of keying material. One is for pairwise key establishment with sensor nodes of its cluster. The other is for pairwise key establishment with other cluster heads. The keying material pre-distribution is illustrated in the Figure 11.5.

Pairwise key establishment

Since the key establishment between base station and cluster heads has been presented in the keying material pre-distribution section, this pairwise key establishment only deals with three scenarios: sensor node to sensor node, sensor node to cluster head, and cluster head to cluster head.

However due to the similarity of the scenarios, we only detail the first scenario between two communicating nodes S_{ik} and S_{jk} in the cluster C_k.

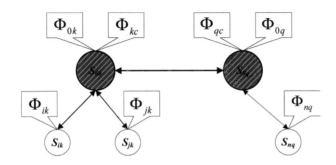

Fig. 11.5 Keying materials in each type of network entities.

The procedure occurs as follows with an added step to allow explicit key authentication.

1. $S_{ik}(S_{jk})$ sends its effective ID y_{ik} and a nonce n_{ik} (y_{jk} and n_{jk}) to the other node. N_i and N_j are used here to provide strong message freshness [30].

$$S_{ik} \rightarrow S_{jk}: y_{ik}|n_{ik}$$
$$S_{jk} \rightarrow S_{ik}: y_{jk}|n_{jk}$$

2. S_{ik} and S_{jk} compute the possible pairwise key $K_{ij}(K_{ji})$ using Φ_{ik}, Φ_{jk}, y_{ik}, and y_{jk} as given in equations (12.5) and (12.6).

3. Up to this step, S_{ik} (S_{jk}) needs to certify that the other has the same key as the one it computed. To do this, S_{ik} (S_{jk}) has to show the other that it has the other's computed key by revealing secret information without revealing the computed key. As in [77], S_{ik} (S_{jk}) generates a message, calculates the message authentication code (MAC) of the message as a function of the message and its computed key and then send the message plus MAC to the receiver (MAC can be calculated using a key-dependent one-way hash function such as HMAC [78]).

$$S_{ik} \rightarrow S_{jk}: y_{ik}|y_{jk}|n_{jk}, MAC(K_{ij}, y_{ik}|y_{jk}|n_{jk})$$
$$S_{jk} \rightarrow S_{ik}: y_{jk}|y_{ik}|n_{jk}, MAC(K_{ji}, y_{jk}|y_{ik}|n_{jk})$$

4. The recipient performs the same calculation on the received message, using its computed key, to generate a new MAC. The received MAC is compared to the calculated MAC. If the received MAC matches the calculated MAC then the receiver is assured that the message is from the alleged sender and its computed key is exactly the same as that of the alleged sender. Since no one else knows the secret key, no one else could prepare a message with a proper MAC.

Performance Analysis

Storage Cost

In our scheme, each sensor node needs to store its keying material and level two broadcast key. Let us assume that the average length of all key types

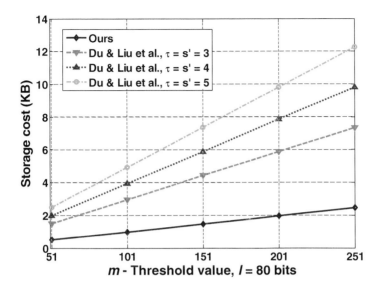

Fig. 11.6 Storage cost of our scheme compared with other schemes.

is l and the average value of all m_is $(i = \overline{1, G})$ is m, then the storage cost for key pre-distribution is $l \times m/8$ byte.

Figure 11.6 compares the storage cost of our scheme with other schemes of the same type [33, 34] with varied values of parameters. As can be seen, with the same values of parameters, our scheme consumes much less memory storage than the compared schemes do. Furthermore, with that storage cost, our scheme always guarantees successful pairwise key establishment while the others do not.

Computational and Communication Overhead

It is easy to realise that our scheme is mainly based on multiplication of matrices $M_\omega (\omega = \overline{1, l})$ and sensor nodes' effective IDs over $GF(2)$. This multiplication essentially is exclusive-OR and AND bit operations. Therefore, it consumes much less computational time and require much less energy as well. Therefore, the overall computational overhead of our scheme is not worth considering. Meanwhile sensor nodes in Du *et al.* scheme [33] and Liu *et al.* scheme [34] need to do multiplication and addition over large finite field which incur much more computational overhead.

In order to establish a pairwise key, only two messages need to be sent and received. These messages easily fit into a packet. Thus the communication overhead of our scheme is very small and comparable to the other schemes.

Security Analysis

Defeating Fake ID Attack

A fake ID attack is a kind of attack in which the attacker captures one or a few nodes to obtain their keying materials, clones the materials onto the attacker's new nodes whose IDs are different from the compromised nodes' IDs, then injects the fake ID nodes into the sensor network. These nodes then try to establish pairwise keys with their neighbors. Once they are accepted by their neighbors, they can launch various insider attacks such as injecting false data packets. However, we claim that such attack is infeasible in our scheme. For ease of presentation, assume that y_c, y_f, y_n are effective IDs of a compromised node S_c, a fake ID node S_f, and S_f 's neighboring node S_n respectively. S_f contains S_c's keying material Φ_c. During the pairwise key establishment, using y_n and Φ_c, S_f derives the pairwise key $K_{cn}(K_{nc})$, whereas S_n derives the pairwise key K_{nf}. These two keys are completely different. Thus, our scheme complete defeats the fake ID attack.

Defeating Threshold Collusion Attack

In this section we argue that our proposed scheme is immune from the node capture based attacks including threshold collusion attack discussed in [76].

As discussed in [76], $M_\omega(\omega = \overline{1,l})$ is a $(m \times m)$ matrix. Using m linearly independent keying material Φ_is, all l M_ωs can be easily revealed. Therefore, m is the value of the collusion threshold. In other words, an adversary only needs to compromise m sensor nodes to be able to compute any pairwise key of any two uncompromised communicating sensor nodes using their effective IDs. It implies that with only m compromised sensor nodes, the adversary can compromise the entire network. In our scheme, since different clusters use different keying materials, the collusion attack could only be mounted locally within each cluster (or among group of cluster heads). Therefore, by selecting the threshold value of m_i for each cluster C_i (or m_c as for group of cluster heads) such that $m_i > N_i(m_c > G)$, the attacker has no way to gather enough m_i linearly independent keying material Φ_{ji}s($j = \overline{0, N_i - 1}$). In other words, the threshold collusion attack in is infeasible to our scheme.

Table 11.3 Comparison of security aspects of major pairwise key establishment schemes.

Scheme property	ours	LEAP+ (Basic) [13]	LEAP+ (Extended) ([13])	RandKPS ([2,17] etc)	ThresRand-KPS ([1, 16] etc)
Resilience to large number of compromised nodes	✓	✓	✓		
Resilience to fake ID attack	✓				✓
Resilience to threshold collusion attack	✓	N/A	N/A	N/A	
Resilience to key-swapping collusion attack			✓		
Resilience to node replication attack			✓		
Guaranteed key establishment	✓				
Direct key establishment	✓				
Resilience to dynamic network topology	✓				

Comparison with Other Schemes

In this section we compare our scheme with several other noteworthy pairwise key establishment schemes with respect to security property. The results are shown in table 3 in which guaranteed key establishment is the property that should guarantee that any two nodes can establish a pairwise key whenever needed. Direct key establishment is the property that requires a scheme to allow two nodes that can communicate directly or indirectly with each other to establish a pairwise key without exposing secrets to or obtaining secrets from any third parties. The table shows that overall, our scheme outperforms other schemes expect the LEAP+ extended version concerning the node replication attack and key-swapping collusion attack. However, due to small size of each cluster, two aforementioned attacks can be easily thwarted by adopting the centralized detection scheme as discussed in [5].

11.5 Summary of Chapter

As distributed sensor networks have attracted enormous research interests due to their extensive and useful applications in military as well as civilian missions, security for them is indispensable especially in adversarial environments. This chapter offered a comprehensive overview of security issues in general and key establishment schemes in particular for DSNs that have been presented in the literature. Moreover, a new deterministic PKE scheme for DSNs is proposed leveraging the common clustering topology to eliminate the security limitations of the existing schemes and avoid security threats with ease. It is worth mentioning that the effort of providing these security features require no added costs to resource-constrained sensor nodes. This is shown via performance analysis with low cost requirements of storage, computation and communication.

References

[1] L. Eschenauer and V. Gligor, "A Key-management scheme for distributed sensor networks," in Proc. 9th ACM Conference on Computer and Communications Security (CSS'02), Washington, DC, USA, 2002, pp. 41–47.

[2] C. Karlof and D. Wagner, "Secure routing in wireless sensor networks: Attacks and countermeasures," in Proc. 1st IEEE Int'l Workshop on Sensor Network Protocols and Applications, 2003, pp. 113–127.

[3] A. D. Wood and J. A. Stankovic, "Denial of service in sensor networks," Computer, vol. 35, pp. 54–62, 2002.

[4] H. Chan, A. Perrig, and D. Song, "Random key predistribution schemes for sensor networks," in Proc. IEEE Symposium on Security and Privacy: IEEE Computer Society, 2003, pp. 197–213.

[5] B. Parno, A. Perrig, and V. Gligor, "Distributed detection of node replication attacks in sensor networks," in Proc. IEEE Symposium on Security and Privacy, 2005, pp. 49–63.

[6] T. Moore, "A collusion attack on pairwise key predistribution schemes for distributed sensor networks," in Proc. 4th Annual IEEE Int'l Conference on Pervasive Computing and Communications (PerCom'06), 2006.

[7] T. T. Dai and J. I. Agbinya, "A framework for confronting key-swapping collusion attack on random pairwise key pre-distribution schemes for distributed sensor networks," in Proc. 5th IEEE Int'l Conference on Mobile Ad Hoc and Sensor Systems (MASS'08), 2008, pp. 815–820.

[8] L. Donggang and D. Qi, "Detecting misused keys in wireless sensor networks," in Proc. IEEE Int'l on Performance, Computing, and Communications Conference (IPCCC'07), 2007, pp. 272–280.

[9] D. W. Carman, P. S. Kruus, and B. J. Matt, "Constraints and Approaches for Distributed Sensor Network Security," DARPA Project report, Cryptographic Technologies Group, Trusted Information System, NAI Labs, vol. 1, Sep. 2000.

[10] R. Roman, C. Alcaraz, and J. Lopez, "A survey of cryptographic primitives and implementations for hardware-constrained sensor network nodes," Mobile Networks and Applications, vol. 12, pp. 231–244, 2007.

[11] Y. W. Law, J. Doumen, and P. Hartel, "Survey and benchmark of block ciphers for wireless sensor networks," ACM Transactions on Sensor Networks (TOSN), vol. 2, pp. 65–93, 2006.

[12] H. Eberle, A. Wander, N. Gura, C.-S. Sheueling, and V. Gupta, "Architectural extensions for elliptic curve cryptography over GF(2ˆm) on 8-bit microprocessors," in Proc. 16th IEEE Int'l Conference on Application-Specific Systems, Architecture Processors (ASAP'05), 2005, pp. 343–349.

[13] K. Sakiyama, L. Batina, B. Preneel, and I. Verbauwhede, "HW/SW Co-design for Accelerating Public-Key Cryptosystems over GF(p) on the 8051 μ-controller," in Proc. World Automation Congress (WAC'06), 2006, pp. 1–6.

[14] A. S. Wander, N. Gura, H. Eberle, V. Gupta, and S. C. Shantz, "Energy analysis of public-key cryptography for wireless sensor networks," in Proc. 3rd IEEE Int'l Conference on Pervasive Computing and Communications (PerCom'05), 2005, pp. 324–328.

[15] E. M. Popovici, "Coding and cryptography for resource constrained wireless sensor networks: A hardware-software co-design approach," in Proc. Int'l Semiconductor Conference, 2006, pp. 19–27.

[16] L. An and N. Peng, "TinyECC: A Configurable Library for Elliptic Curve Cryptography in Wireless Sensor Networks," in Proc. 7th Int'l Conference on Information Processing in Sensor Networks: IEEE Computer Society, 2008.

[17] W. Ronald, K. Derrick, C. Sue-fen, G. Charles, L. Charles, and K. Peter, "TinyPK: securing sensor networks with public key technology," in Proc. 2nd ACM workshop on Security of Ad hoc and Sensor Networks Washington DC, USA: ACM, 2004.

[18] P. Steffen, L. Peter, and P. Krzysztof, "Public key cryptography empowered smart dust is affordable," Int'l Journal on Sensor Network, vol. 4, pp. 130–143, 2008.

[19] G. Gaubatz, J. P. Kaps, and B. Sunar, "Public key cryptography in sensor networks-revisited," in Proc. 1st European Workshop Security in Ad-Hoc and Sensor Networks, 2004, pp. 2–18.

[20] N. Gura, A. Patel, A. Wander, H. Eberle, and S. C. Shantz, "Comparing Elliptic Curve Cryptography and RSA on 8-bit CPUs," Lecture Notes in Computer Science, pp. 119–132, 2004.

[21] G. Prasanth, V. Ramnath, P. Pushkin, D. Alexander, M. Frank, and S. Mihail, "Analyzing and modeling encryption overhead for sensor network nodes," in Proc. 2nd ACM Int'l Conference on Wireless Sensor Networks and Applications San Diego, CA, USA: ACM, 2003.

[22] M. Eltoweissy, M. Moharrum, and R. Mukkamala, "Dynamic key management in sensor networks," IEEE Communications Magazine, vol. 44, pp. 122–130, 2006.

[23] R. Anderson, H. Chan, and A. Perrig, "Key Infection: Smart Trust for Smart Dust," in Proc. 12th IEEE Int'l Conference on Network Protocols (ICNP'04), 2004, pp. 206–215.

[24] C. Siu-Ping, R. Poovendran, and S. Ming-Ting, "A key management scheme in distributed sensor networks using attack probabilities," in Proc. IEEE Global Telecommunications Conference (GLOBECOM'05), 2005, p. 5.

[25] H. Dijiang, M. Manish, M. Deep, and H. Lein, "Location-aware key management scheme for wireless sensor networks," in Proc. 2nd ACM Workshop on Security of Ad hoc and Sensor Networks Washington DC, USA: ACM, 2004.

[26] W. Du, J. Deng, Y. S. Han, S. Chen, and P. K. Varshney, "A Key Management Scheme for Wireless Sensor Networks Using Deployment Knowledge," in Proc. IEEE INFOCOM, 2004, pp. 586–597.

[27] A. Farooq, "Location dependent key management using random key-predistribution in sensor networks," in Proc. 5th ACM Workshop on Wireless Security Los Angeles, California: ACM, 2006.

[28] D. Liu and P. Ning, "Improving key predistribution with deployment knowledge in static sensor networks," ACM Trans. on Sensor Networks, vol. 1, pp. 204–239, 2005.

[29] S. Zhu, S. Setia, and S. Jajodia, "LEAP+: Efficient Security Mechanisms for Large-Scale Distributed Sensor Networks," ACM Trans. on Sensor Networks, vol. 2, pp. 500–528, 2006.

[30] A. Perrig, R. Szewczyk, J. D. Tygar, V. Wen, and D. E. Culler, "SPINS: security protocols for sensor networks," Wireless Network, vol. 8, pp. 521–534, 2002.

[31] M. F. Younis, K. Ghumman, and M. Eltoweissy, "Location-aware combinatorial key management scheme for clustered sensor networks," IEEE Trans. on Parallel and Distributed Systems, vol. 17, pp. 865–882, 2006.

[32] T. T. Dai and C. S. Hong, "Efficient ID-based threshold random key pre-distribution scheme for wireless sensor networks," IEICE Trans. on Communications, vol. E91-B, pp. 2602–2609, Aug. 2008.

[33] W. Du, J. Deng, Y. S. Han, P. K. Varshney, J. Katz, and A. Khalili, "A Pairwise Key Pre-distribution Scheme for Wireless Sensor Networks," ACM Trans. on Information & System Security, vol. 8, pp. 228–258, 2005.

[34] D. Liu, P. Ning, and R. Li, "Establishing Pairwise Keys in Distributed Sensor Networks," ACM Trans on Information & System Security, vol. 8, pp. 41–77, 2005.

[35] W. Zhang, M. Tran, S. Zhu, and G. Cao, "A random perturbation-based scheme for pairwise key establishment in sensor networks," in Proc. 8th ACM Int'l Symposium on Mobile Ad hoc Networking and Computing Montreal, Quebec, Canada: ACM, 2007.

[36] W. Zhang and G. Cao, "Group rekeying for filtering false data in sensor networks: A predistribution and local collaboration-based approach," in Proc. IEEE INFOCOM, 2005, pp. 503–514.

[37] J. Brown, D. Xiaojiang, and K. Nygard, "An Efficient Public-Key-Based Heterogeneous Sensor Network Key Distribution Scheme," in Proc. IEEE Global Telecommunications Conference (GLOBECOM'07), 2007, pp. 991–995.

[38] G. Gaubatz, J. P. Kaps, E. Ozturk, and B. Sunar, "State of the art in ultra-low power public key cryptography for wireless sensor networks," in Proc. 3rd IEEE Int'l Conference on Pervasive Computing and Communications Workshops (PerCom'05), 2005, pp. 146–150.

[39] D. J. Malan, M. Welsh, and M. D. Smith, "A public-key infrastructure for key distribution in TinyOS based on elliptic curve cryptography," in Proc. 1st Annual IEEE Communications Society Conference on Sensor and Ad Hoc Communications and Networks (SECON'04), 2004, pp. 71–80.

[40] J. Qi, H. Jianbin, and C. Zhong, "C4W: An energy efficient public key cryptosystem for large-scale wireless sensor networks," in Proc. IEEE Int'l Conference on Mobile Adhoc and Sensor Systems (MASS'06) 2006, pp. 827–832.

[41] A. V. S. J. M. Alfred, and C. V. O. Paul, Handbook of Applied Cryptography: CRC Press, Inc., 1996.

[42] C. Haowen, V. D. Gligor, A. Perrig, and G. Muralidharan, "On the distribution and revocation of cryptographic keys in sensor networks," IEEE Trans on Dependable and Secure Computing, vol. 2, pp. 233–247, 2005.

[43] J. N. Al-Karaki and A. E. Kamal, "Routing techniques in wireless sensor networks: A survey," IEEE Wireless Communications, vol. 11, pp. 6-28, 2004.

[44] J. Deng, R. Han, and S. Mishra, "A Performance Evaluation of Intrusion-Tolerant Routing in Wireless Sensor Networks," in Proc. Int'l Conference on Information Processing in Sensor Networks (IPSN'03), 2003, pp. 349–364.

[45] R. Di Pietro, L. V. Mancini, L. Yee Wei, S. Etalle, and P. Havinga, "LKHW: a directed diffusion-based secure multicast scheme for wireless sensor networks," in Proc. Int'l Conference on Parallel Processing Workshops, 2003, pp. 397–406.

[46] T. Malik, Y. Jian, P. Biswajit, and M. Sanjay, "A secure hierarchical model for sensor network," SIGMOD Rec., vol. 33, pp. 7–13, 2004.

[47] D. Liu, P. Ning, S. Zhu, and S. Jajodia, "Practical broadcast authentication in sensor networks," in Proc. 2nd Annual Int'l Conference on Mobile and Ubiquitous Systems: Networking and Services (MobiQuitous'05), 2005, pp. 118–129.

[48] W. Du, J. Deng, Y. S. Han, and P. K. Varshney, "A witness-based approach for data fusion assurance in wireless sensor networks," in Proc. IEEE Global Telecommunications Conference (GLOBECOM'03), 2003, pp. 1435–1439.

[49] Z. Yanchao, L. Wei, L. Wenjing, and F. Yuguang, "Location-based compromise-tolerant security mechanisms for wireless sensor networks," IEEE Journal on Selected Areas in Communications, vol. 24, pp. 247–260, 2006.

[50] Y. Fan, H. Luo, L. Songwu, and Z. Lixia, "Statistical en-route filtering of injected false data in sensor networks," IEEE Journal on Selected Areas in Communications, vol. 23, pp. 839–850, 2005.

[51] Z. Sencun, S. Sanjeev, J. Sushil, and N. Peng, "Interleaved hop-by-hop authentication against false data injection attacks in sensor networks," ACM Trans. on Sensor Networks, vol. 3, p. 14, 2007.

[52] I. Chalermek, G. Ramesh, and E. Deborah, "Directed diffusion: A scalable and robust communication paradigm for sensor networks," in Proc. 6th Annual Int'l Conference on Mobile Computing and Networking Boston, Massachusetts, United States: ACM, 2000.

[53] C. Castelluccia, E. Mykletun, and G. Tsudik, "Efficient aggregation of encrypted data in wireless sensor networks," in Proc. 2nd Annual Int'l Conference on Mobile and Ubiquitous Systems: Networking and Services (MobiQuitous'05), 2005, pp. 109–117.

[54] C. Jen-Yeu, P. Gopal, and X. Dongyan, "Robust computation of aggregates in wireless sensor networks: distributed randomized algorithms and analysis," in Proc. 4th Int'l Symposium on Information Processing in Sensor Networks (IPSN'05) Los Angeles, California: IEEE Press, 2005.

[55] W. David, "Resilient aggregation in sensor networks," in Proc. 2nd ACM Workshop on Security of Ad hoc and Sensor Networks Washington DC, USA: ACM, 2004.

[56] S. Naveen, S. Umesh, and W. David, "Secure verification of location claims," in Proc. 2nd ACM Workshop on Wireless Security San Diego, CA, USA: ACM, 2003.

[57] L. Loukas and P. Radha, "SeRLoc: Secure range-independent localization for wireless sensor networks," in Proc. 3rd ACM Workshop on Wireless Security Philadelphia, PA, USA: ACM, 2004.

[58] S. Capkun and J. P. Hubaux, "Secure positioning of wireless devices with application to sensor networks," in Proc. 24th Annual Joint Conference of the IEEE Computer and Communications Societies (INFOCOM'05), 2005, pp. 1917–1928.

[59] Z. Yanchao, L. Wei, and F. Yugang, "Secure localization in wireless sensor networks," in Proc. IEEE Military Communications Conference (MILCOM'05), 2005, pp. 3169–3175 Vol. 3165.

[60] Z. Yanchao, L. Wei, F. Yuguang, and W. Dapeng, "Secure localization and authentication in ultra-wideband sensor networks," IEEE Journal on Selected Areas in Communications, vol. 24, pp. 829–835, 2006.

[61] O. Celal, Z. Yanyong, and T. Wade, "Source-location privacy in energy-constrained sensor network routing," in Proc. 2nd ACM Workshop on Security of Ad hoc and Sensor Networks Washington DC, USA: ACM, 2004.

[62] P. Kamat, Z. Yanyong, W. Trappe, and C. Ozturk, "Enhancing source-location privacy in sensor network routing," in Proc. 25th IEEE Int'l Conference on Distributed Computing Systems (ICDCS'05), 2005, pp. 599–608.

[63] W. Yong, G. Attebury, and B. Ramamurthy, "A survey of security issues in wireless sensor networks," IEEE Communications Surveys & Tutorials, vol. 8, pp. 2–23, 2006.

[64] S. Bo, L. Osborne, X. Yang, and S. Guizani, "Intrusion detection techniques in mobile ad hoc and wireless sensor networks," IEEE Wireless Communications, vol. 14, pp. 56–63, 2007.

[65] R. d. S. Ana Paula, H. T. M. Marcelo, P. S. R. Bruno, A. F. L. Antonio, B. R. Linnyer, and W. Hao Chi, "Decentralized intrusion detection in wireless sensor networks," in Proc. 1st ACM Int'l Workshop on Quality of Service & Security in Wireless and Mobile Networks Montreal, Quebec, Canada: ACM, 2005.

[66] I. Onat and A. Miri, "An intrusion detection system for wireless sensor networks," in Proc. IEEE Int'l Conference on Wireless And Mobile Computing, Networking And Communications (WiMob'05), 2005, pp. 253–259 Vol. 253.

[67] S. Chien-Chung, C. Ko-Ming, K. Yau-Hwang, and H. Mong-Fong, "The new intrusion prevention and detection approaches for clustering-based sensor networks [wireless sensor networks]," in Proc. IEEE Wireless Communications and Networking Conference (WCNC'05), 2005, pp. 1927–1932.

[68] D. Martynov, J. Roman, S. Vaidya, and F. Huirong, "Design and implementation of an intrusion detection system for wireless sensor networks," in Proc. IEEE Int'l Conference on Electro/Information Technology, 2007, pp. 507–512.

[69] C. Sheng-Tzong, Y. Szu, and C. Chia-Mei, "Distributed Detection in Wireless Sensor Networks," in Proc. 7th IEEE/ACIS Int'l Conference on Computer and Information Science (ICIS'08), 2008, pp. 401–406.

[70] J. Spencer, The Strange Logic of Random Graphs vol. 22: Springer-Verlag, 2000.

[71] B. Carlo, S. Alfredo De, V. Ugo, H. Amir, K. Shay, and Y. Moti, "Perfectly secure key distribution for dynamic conferences," Information Computing, vol. 146, pp. 1–23, 1998.

[72] L. Gong and D. J. Wheeler, "A matrix key-distribution scheme," Journal of Cryptology, vol. 2, pp. 51–59, 1990.

[73] R. Blom, "An optimal class of symmetric key generation systems," in Proc. EUROCRYPT 84 Workshop on Advances in Cryptology: Theory and Application of Cryptographic Techniques Paris, France: Springer-Verlag New York, Inc., 1985.

[74] H. Fu, S. Kawamura, M. Zhang, and L. Zhang, "Replication Attack on Random Key Predistribution Schemes for Wireless Sensor Networks," Computer Communication, vol. 31, pp. 842–857, 2008.

[75] C. Mauro, P. Roberto Di, M. Luigi Vincenzo, and M. Alessandro, "A randomized, efficient, and distributed protocol for the detection of node replication attacks in wireless sensor networks," in Proc. 8th ACM Int'l Symposium on Mobile Ad hoc Networking and Computing Montreal, Quebec, Canada: ACM, 2007.

[76] M. Tsutomu and I. Hideki, "On the Key Predistribution System: A Practical Solution to the Key Distribution Problem," in A Conference on the Theory and Applications of Cryptographic Techniques on Advances in Cryptology: Springer-Verlag, 1988.

[77] W. Stallings, Cryptography And Network Security: Principles and Practice: Prentice Hall, 2006.

[78] M. Y. Rhee, Internet Security: Cryptographic Principles, Algorithms and Protocols: Wiley, 2003.

12

Planning and Addressing of Wireless Sensor Networks

Johnson Ihyeh Agbinya

University of Technology Sydney, Australia

12.1 Introduction

Sensor networks need to be planned but unfortunately most reported implementations are ad hoc in nature and hence lack the required robust performance. A great deal of current research on wireless sensor networks is dominated by reliance on models and experiments based on random distribution of sensors in the network and within the environments without a systematic network planning. In practice, sensor networks that will be used for day to day applications will need to be planned for coverage, performance and throughput. Understandably, the lack of planning methods for sensor networks is a limiting factor against practical implementations. Understanding how to plan sensor networks will lead to significant implementations and applications. The major objectives for planning sensor networks include enhancement of connectivity, better area of coverage, reduced interference, optimum monitoring or sensing of conditions within the geographical region of interest, topology control and network management. Unplanned sensor networks therefore effectively work against the desired objectives. Optimal placement of sensors reduces the costs of network ownership and management and enhances the detection potential of the network. A reusable sensor network planning method provides a template for rapid deployment of sensor networks in different environments and for different applications. A planned sensor network is a multi-criteria optimisation problem which involves minimization of the number of sensors, minimum energy consumption, maximum range of coverage, optimum connectivity, best placement and optimum sensing. These requirements are usually not complementary. While the sensor placement problem seeks to understand how the least number of sensors can be placed in an environment to achieve optimum sensing (high connectivity and coverage) at reduced cost. The sensitivity of the sensor determines the coverage range and the coverage range determines the network connectivity. Hence the available power to the sensor affects its coverage range.

This chapter therefore addresses the problems of where and how best to place the sensors for optimum tracking of mobile objects. It also addresses the issues of sensor selection for best coverage in addition to optimising the coverage area. We intentionally take a pragmatic approach and assume that it is possible through good engineering to optimally place sensors to achieve an objective. Sensors do not radiate omni-directionally as is assumed in most papers on sensor networks [1]. In fact for example PIR sensors are highly directive and have limited fields of views.

Unplanned tracking wireless sensor networks for surveillance focus generally on ability to detect motion but miss the essential features of best coverage and range. By optimally placing the sensors, choosing appropriate sensors with the required field of view and with power and frequency planning the capabilities of very short coverage range sensors can be maximised and the scalability of such networks becomes significantly easier. We have studied extensively the capabilities of passive infra red sensors. Passive infra red (PIR) sensors are typically used for motion detection over very short ranges and subtend very

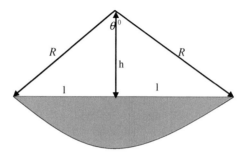

Fig. 12.1 Node S Field of View Using PIR Sensor.

small angles over which they can detect motion. Figure 12.1 is one example. In the Figure, a sensor of range R and field of view θ is located at the vertex of the triangle. The height of the triangle is h and based on the parameters shown it is possible to plan a sensor network with optimum area coverage.

The following basic assumptions were made. We assume it is possible to create a template that permits reuse. We also assume that the field of view of the sensor is known and the area of coverage is defined. Therefore, the length, height and area of each of the two triangles in the Figure are known and given by the expressions:

$$l = R \sin \left(\frac{\theta}{2} \right) \quad and \quad h = R \cos \left(\frac{\theta}{2} \right) \tag{12.1}$$

Therefore the area of each of the two triangles is, $a = lh/2 = 1/2R^2 \cos(\theta/2) \sin(\theta/2)$. The area of the purple region in Figure 12.1 can be shown to be:

$$a_p = R^2 \left(\theta - \sin \left(\frac{\theta}{2} \right) \cos \left(\frac{\theta}{2} \right) \right) = \frac{1}{2} R^2 (2\theta - \sin \theta) \tag{12.2}$$

This area is significantly larger than the area of the large triangle subtended by the angle θ (radians). The largest coverage distance is obtained along the base of this triangle and is significantly larger than the range R.

12.1.1 Planning of a Wireless Sensor Chain Network

Define the term node S to refer to the sensor and the view angle as the cell. A linear sensor network can be formed with several node S arranged in an array and used to monitor a highway or street with sensors deployed on light poles where suitable or on installed fabric such as poles. In public buildings and factories, the sensors could be suspended inside from the ceiling or on roofs. They can also be hidden inside embedded lighting structures in the roof. In a liner chain network of sensors with field of view $(\theta, L = 2l)$, the number of sensors required to cover a distance D metres is therefore given by the expression:

$$N_c = \frac{D}{2l} = \frac{D}{2R \sin \left(\frac{\theta}{2} \right)} \tag{12.3}$$

This is the optimum distance to be covered and also the least number of sensors required to ensure that every point on the line is exposed by at least one sensor.

The area covered by each of the sensors is maximum along the base of the big triangle and is given by the expression

$$a_s = \pi.r^2 = \pi l^2 = \pi R^2 \sin^2\left(\frac{\theta}{2}\right) \qquad (12.4)$$

The total spot area exposed by the sensors is therefore given by the expression:

$$N_c a_s = N_c \pi.r^2 = N_c \pi l^2 = \frac{\pi D R \sin\left(\frac{\theta}{2}\right)}{2} \qquad (12.5)$$

Consider the application of a liner array of network of sensors that is used for tracking of mobile objects within an open space. The distance over which tracking takes place is D_t. Each PIR tracks within a distance of 1 metres. Therefore the number S of PIR sensors required is:

$$S = \frac{D_T}{2l} = \frac{D_T}{2R \sin\left(\frac{\theta}{2}\right)} \qquad (12.6)$$

Assume that q PIR sensors can be mounted on a single target board. Therefore the total number of target boards T required to support the tracking application is:

$$T = \frac{S}{q} \qquad (12.7)$$

Since each PIR sensor provides also a spot coverage area, the total ground exposure of the linear array is given by the expression

$$A = S.\pi l^2 = \pi \left(R\sin\left(\frac{\theta}{2}\right)\right)^2 \frac{D_T}{2R\sin\left(\frac{\theta}{2}\right)} \qquad (12.8)$$

$$= \frac{\pi D_T R \sin\left(\frac{\theta}{2}\right)}{2} \; square\_metres$$

The sensors are suspended and look down at the ground.

Planning of a Wide Area Wireless Sensor Network

When the sensors are installed to look horizontally as in Figures 12.2 and 12.3 optimum range tracking can take place along the line linking the base of the view triangles. Since the base length of each triangle is 2l, for S sensors the total edge-to-edge tracking distance is 2Sl or:

$$Track = 2Sl = 2S.R\sin\left(\frac{\theta}{2}\right) \qquad (12.9)$$

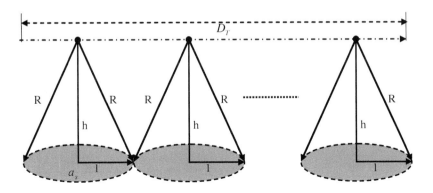

Fig. 12.2 Total area of Coverage of S PIR Sensors.

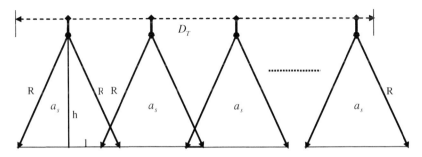

Fig. 12.3 Intersecting Triangular Sensor Network Cells.

The total view area for the S sensors is also given by the expression:

$$A_S = Sa_s = \frac{\pi D_T R}{2} \sin\left(\frac{\theta}{2}\right) \qquad (12.10)$$

We choose to use Figure 12.3 as a template for the case when the fields of view are concatenated edge to edge resulting to Figure 12.4. This process can be repeated to achieve tracking over wide area with more sensors. The inside of each quadrilateral is covered by at least two sensors and at most four if the sensors are deployed at the vertices of the quadrilateral. The inner area of each quadrilateral is

$$A_q = 2a_s = 2\pi R^2 \sin^2\left(\frac{\theta}{2}\right) \qquad (12.11)$$

In Figure 12.4, each sensor creates a triangular cell which can be re-used as a planning template for a wide area sensor network.

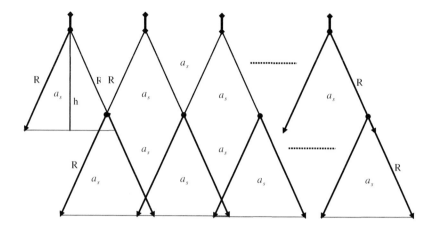

Fig. 12.4 Wide area Sensor Network Using Triangular Node S Cells.

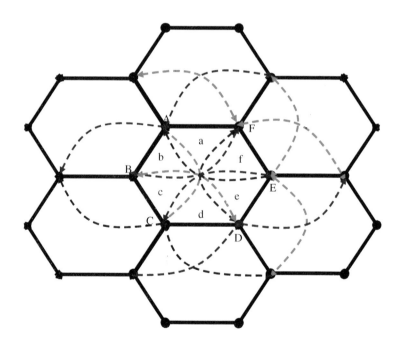

Fig. 12.5 Hexagonal Sensor Cells Formed with 6 Node S Per Hexagon.

In Figure 12.5, the central hexagon is edge excited by 6 base station sensors (A, B, C, D, E and F) with each having 3 sectors. One sector in each base station faces in the inside are of the central hexagon. This central hexagon is segmented into 13 regions. The following deductions can be made with respect to the 13 regions:

i) In the hexagon, motion at every point is detectable by at least two sensors

ii) Motion at the centre point of the hexagon is detectable by six different sensors simultaneously and the size of this 'point' is equal to the vertical projection of the object onto the plane of the hexagon. The probability of the object being tracked lying within this region is:

$$p_1 = \frac{a_1}{A_h}$$

Where A_h and a_1 are the areas of the region and hexagon respectively.

iii) There are six sub-regions (a, b, c, d, e and f) within the hexagon where motion is detectable by two sensors simultaneously and the probability of an object lying in those regions is given by the expression

$$p_2 = \frac{a_2}{A_h} \tag{12.12}$$

a_2 is the total area of the regions. The sensors picked by a mobile object within these regions are given in Table 12.1. Assume the six sensors are connected to one target board.

iv) There are six sub-regions within the hexagon where motion is detectable by three sensors simultaneously and the probability of

Table 12.1 Motion Detection with Hexagonal Sensor Cells.

Region	Node A	Node B	Node C	Node D	Node E	Node F	Coordinator Reading	# of sensors detecting motion
	2^1	2^3	2^5	2^0	2^2	2^4		
a	1	0	0	0	0	1	18	2
b	1	1	0	0	0	0	10	2
c	0	1	1	0	0	0	40	2
d	0	0	1	1	0	0	33	2
e	0	0	0	1	1	0	5	2
f	0	0	0	0	1	1	20	2

an object lying in those regions is given by the expression

$$p_3 = \frac{a_3}{A_h} \qquad (12.13)$$

a_3 is the total area of the regions. An object is within this region if it also lies on any of the arcs that demarcate these sub-regions. In general, the following inequality holds: $a_1 << a_3 \leq a_2$. By labelling or addressing the sensors appropriately we can determine which sensors are detecting motion and also where within the hexagon the object is.

The objective is to create a wide area land mine detection region. The PIR sensors have each an angular view of 140 degrees which we map into three sectors of 120 degrees to form a circular view. The PIR coverage of two sensors intersects by 20 degrees around each edge line of the hexagon. In this 20 degrees region two sensors have good view of a moving object. The maximum range extent of each PIR lies on an arc.

The perimeter of the arc in Figure 12.6 provides the longest tracking curved distance with this PIR and is equal to $R.\theta$ metres. Clearly the field of view provides an indicator of how long the distance over which a sensor can detect motion. The bigger the angle of view the longer the distance over which motion is detectable by the sensor. Therefore by sectoring the sensor base station, its field of view is also limited and an omnidirectional sensor base station is preferable when large motion detection coverage is required. Unfortunately omnidirectional antennas also interfere more with other neighbouring sensors and hence the choice of the PIR should be made carefully based on the motion detection application.

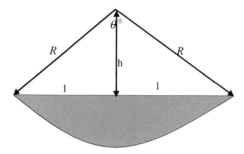

Fig. 12.6 Tracking Circular Arc of a PIR Sensor.

12.2 Implementation

We have implemented the planning scenario in Figures 12.1–12.5. The base sensor monitoring kit consists of a Texas Instruments ez430-RF2480 series which operate as Zigbee Transceiver. It was factory fabricated to read temperatures using its target boards. The target boards are battery powered using two AAA batteries. The eZ430-RF2480 operates as a sensor monitoring board (Coordinator) which plugs into a computer through a USB connection. It has a line of sight range of about 30 to 40m in clear space. It coordinates the functions of stand alone target boards operating over a frequency of 2.4GHz. Target boards can be configured to operate as Routers. In this mode they permit multi hopping from each other. They can also be configured as end devices in which case they have reduced functions compared with the routers and coordinators. Each target board has provision for five sensors of any type to be connected at the same time. In our case we are interested in motion detection and therefore interfaced passive infrared (PIR) sensors. Each ez-430-RF2480 board works off a 3 volts battery with low power drainage.

For motion detection we used the PIR sensors as end devices. Three PIRs were connected to one target board and shared a 9 volts power source, although they are individually able to operate over a maximum voltage of 24 volts. Measured range coverage of the sensors is a function of the input power and by sharing a single 9 volt battery the line of sight coverage of each PIR was limited to about 3 metres.

The three motion sensors were connected so that the outputs can be read as a three bit number (as in Table 12.2) of the form: 2^1 2^0 2^2. Hence we were able to tell which sensor was detecting motion.

Table 12.2 Sensor ID Logic.

2^1	2^0	2^2	Coordinator Reading	# of sensors detecting motion
0	0	0	0	0
0	0	1	1	1
0	1	0	2	1
1	0	0	4	1
0	1	1	3	2
1	0	1	5	2
1	1	0	6	2
1	1	1	7	3

An interface program was written in Matlab which allows the above readings to be captured and saved into log files over extended tracking periods. These readings are plotted to provide tracking profiles. The program is also used to refresh and configure the PIR sensors and the target boards to activate the sensors to display readings.

Several tracking scenarios were investigated. Many situations require a chain of motion sensors in a chain network. A chain network is formed by arranging several sensors so that the coverage of two neighbouring sensors overlap.

A second tracking scenario was investigated. For this case, the motion sensors were mounted on a conical shape so that the three motion sensors form three sectors of three overlapping 140 degrees. Hence tracking around a circular perimeter can be done. Two neighbouring sensors overlap by 20 degrees as in Figure 12.7. Hence within the overlap region two sensors can detect the motion of a single object simultaneously, thus providing motion sensing redundancy. This arrangement was found to provide an advantage over the linear arrangement and hence was replicated in a hexagonal motion sensing structure. At each vertex of the hexagon, three sensors are mounted and point into three sectors as shown in Figure 12.7. Thus the hexagon in the middle has 6 sensors being used to sense motion. Section 12.3 is a presentation on how to address individual sensors in sensor networks.

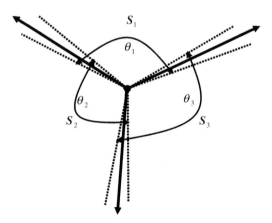

Fig. 12.7 Three sectored Sensor Base Station with Overlapping Fields of Views.

12.3 Address Allocation Schemes and Planning of Sensor Networks

This section is a summary of addressing schemes for personal area and sensor networks based on Zigbee. It discusses addressing schemes based on arithmetic series which avoids address duplication and wastage. Problems associated with skip distance addressing and static addressing schemes are discussed. Progressive addressing of nodes is demonstrated. This permits identifying the Zigbee network using IPv6 addressing of a parent gateway and offset addressing of children nodes.

Zigbee defines two classes of devices the so-called full function devices (FFD) and the reduced function devices (RFD). FFD are expected to be more powerful than the RFD in the network and remain on for longer periods compared with the RFDs. They need to be on because they control their RFDs during communications. Hence while the RFDs may go to sleep when not active, the FFDs need to be awake.

Traditionally, in a Zigbee network, devices (nodes) may assume any of three possible roles, either as coordinator, router or end device. The end devices are barren nodes and do not have child nodes and therefore are limited to executing applications embedded in them. Hence they are either FFD or RFDs. Routers have parent nodes, possess child nodes and also relay messages. They are typically FFDs. Coordinator is an FFD node. It controls a Zigbee network and may act also as a gateway to the external world, an access authenticator and also a trust centre. These functions are inherent in the coordinator. In Figure 12.2, node one (1) is a coordinator, the blue nodes are routers and the brown nodes are end nodes. As shown, there is only one coordinator, many routers and end nodes. Zigbee typically supports multi hopping and may be used in at least three network configurations: star, mesh or tree (Figure 12.2).

12.4 Addressing Schemes in Zigbee Networks

Several device addressing schemes have been published for Zigbee-based wireless networks with the objectives of optimum routing and minimum delay regimes.

12.5 Skip Distance Addressing

A tree routing addressing scheme was proposed [2] for Zigbee net-works. The scheme consists of two components: tree node addresses and tree-address-based routing. The addressing algorithm is controlled by the coordinator and to start the addressing, it assigns itself the root address 0 and a network depth (d) with $d_0 = 0$. The network depth is defined as the number of hops separating the coordinator from the farthest child node in the network. When a child node i wants to associate itself with node k (parent), the parent assigns to the child node an address that is offset from its own address by a value that is a function of the network depth. This offset is called the skip distance. The child therefore is located at a network depth $d_i = d_k + 1$. In the binary tree shown in Figure 12.8, the depth to the child (brown) nodes is $d = 6$. The largest number of child nodes which can associate with a parent is C_m and the maximum number of routers in the network is R_m. For a network with maximum depth L_m, the address a parent node assigns to a RFD child node (n) is:

$$A_n = A_k + C_{skip}(d).x R_m + n \qquad (12.14)$$

If the child node is an FFD it is assigned the address:

$$A_n = A_k + 1 + C_{skip}(d).(n - 1) \qquad (12.15)$$

The skip distance is computed with the expression

$$C_{skip}(d) = \begin{cases} 1 + C_m.(L_m - d - 1), & if \ R_m = 1 \\ \dfrac{1 + C_m - R_m - C_m.R_m^{L_m-d-1}}{1 - R_m}, & otherwise \end{cases} \qquad (12.16)$$

C_{skip} represents the address block that a router at depth d can allocate for the new devices when they request association. When $C_{skip} = 0$ then all addresses have been exhausted and a request for new association is rejected. To route messages or data to a destination, the router uses the address of the destination and its network depth. The authors also show that for a destination D with address A and depth d, if the receiving node is a child then $A < D < A + C_{skip}(d - 1)$. If the destination node is not a child, the router computes the next hop as D (It is necessary to observe the error in equation (17) which originates from [1] and has not been corrected in any literature on the subject) because when the node is not an end device N = D!

$$N = \begin{cases} D, & if\ end\ device \\ A + 1 + \left[\dfrac{D - (A + 1)}{C_{skip}(d)} \right] x C_{skip}(d), & otherwise \end{cases} \quad (12.17)$$

This algorithm is called the skip short address routing mechanism for Zigbee network. Several deficiencies of this addressing and routing scheme are apparent. Since Lm, Cm and Rm, are predetermined before the whole network is designed, the scalability of the network tree is restricted. More deficiencies of the method have been pointed out by Anurag et al [3]. The success of the scheme in [2] is based on the formation of a tree topology. When this fails, addresses are wasted. The scheme also results to a static address allocation and hence in many situations addresses could lie unused in some nodes even when other nodes have exhausted their addresses and are in need of more. The addressing scheme also limits the number of children nodes that a router can permit access. Since the algorithm also assumes a worst case scenario whereby all possible nodes need to have address provision, this leads to a restrictive network depth. Address wastage problem in the skip distance algorithm is described in [3] and [4]. The scheme also limits device mobility [4] and due to the limitation of maximum number of children nodes, significant overhead traffic when many nodes are located within the radio range of a coordinator node. The scheme however leads to optimum routing and provides minimum latency. Hence the scheme can lead to significant quality of service benefits from the reduced routing delays.

12.6 Static Addressing Algorithm

The static addressing algorithm was proposed by Anurag [3] with the objective of solving some of the problems associated with the skip distance method. Significantly, it seeks to prevent the wastage of address space. In this scheme, the maximum number of end devices (E_n) that each router can handle is specified apriori. Then a simple algorithm is used to specify the addresses of the routers. The algorithm is a 'last router, first address' mechanism whereby the router deepest in the network is assigned an address first. Thereafter addresses are assigned by climbing up (Figure 12.8) the tree topology. Each router in the network maintains an address list (the addresses of its children routers). Therefore unlike the skip distance method which allocates addresses for nodes

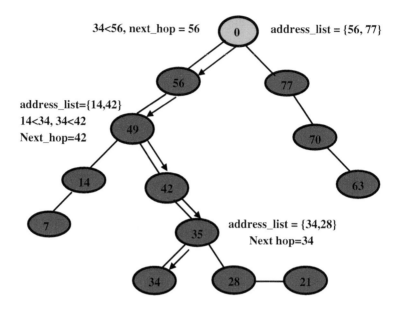

Fig. 12.8 The path of a packet destined to "34" from "0" [3].

not yet in the network, the static address algorithm (SAA) does not. However, it suffers from the network scalability problem due to fixing the number of end devices. The addressing is static.

12.7   Address Allocation Based On Last Address Assigned

Due to the limitation of the skip distance addressing scheme, several modification of the scheme have been reported. One of such modification is the address allocation based on the so-called "last address assigned" [4].

12.8   A New Zigbee Network Addressing Scheme

In this new addressing scheme, we employ the concepts of designating nodes as coordinator, router and end device and also specify the depth of the network in terms of the number of hops required to get to the destination or the last node in the network [5]. These concepts are illustrated in Figure 12.9. In Figure 12.10, the devices are allocated to nodes and addressed from the root node (node 1) to the last. Address 0 is reserved for future use to designate

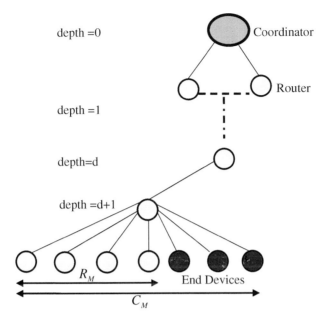

Fig. 12.9 Naming Convention in a Zigbee Tree Network.

the global sensor network node, or what we call the Interpanet (International personal area networking). One feature which has often been left out in current literature is how to determine the number of the nodes at each depth of the network automatically. Our addressing scheme solves this problem. It also solves the problems of address wastage and address conflicts inherent in the addressing schemes described in [2, 3, 4].

Tier1 node addresses use a defined plan based on arithmetic series in the branches [5] and the sum of nodes at each depth (d) based on the sum of the coefficients of the binomial theorem of two variables a and b raised to power d. This means at depth(0) the sum of nodes is 1; at depth(1), the sum is 3; at depth(2), the sum is 7 and at depth(3) the sum is 15 etc. The binomial theorem is well known in basic algebra. This method permits efficient routing (based on arithmetic series, to be shown shortly) and knowing the number of Zigbee nodes at each level of the network. Should there be the need to scale the network, the method about to be described can be re-initialised from a known starting address at each node of the tree. Note that although we use the binomial sum in this chapter for convenience and ease of tracking the nodes locations in the network depth, other series can be used. All the nodes located to the left half of the tree network have addresses defined by the expression:

$$Addresses_{left} = A_k + 2a_{k-1}; \quad where \; A_1 = 1; \quad and \quad a_1 = 2 \quad (12.18)$$

The addresses for the Zigbee nodes to the right half of the tree network starting from the Coordinator are also given by the expression:

$$Addresses_{right} = B_k + 2b_{k-1}; \quad where \; B_1 = 1; \quad and \quad b_1 = 1 \quad (12.19)$$

The skip addresses forming the weights of the links between any two linked nodes are given by the terms of the arithmetic series being used.

Figure 12.10 describes this addressing scheme in a nutshell for tier1 address allocation. In tier1, addresses are allocated to nodes up to the depth 'd' chosen to match the size of the network. The numbers inside the circles in the Figure are the addresses of the nodes. The numbers on the links (link lengths which are equivalent to the incremental terms of our arithmetic series) are used for incremental allocation of addresses down the tree. Observe that for the left branch of the tree at each node, the incremental addresses are twice the value of the increments at the right branch at each node. Odd addresses are

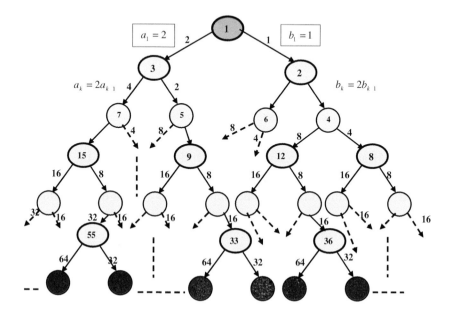

Fig. 12.10 Series-based addressing of Zigbee Network.

Table 12.3 Assigned Addressing Scheme in the Tree.

			(0)				
			{1}				
			(1)				
			{3 2}				
		(3)		(2)			
		{7 5}		{6 4}			
	(7)	(5)		(6)	(4)		
	{15 11}	{13 9}		{13 10}	{12 8}		
(15)	(11)	(13)	(9)	(14)	(10)	(12)	(10)
{31 23}	{27 19}	{29 21}	{25 17}	{30 22}	{26 18}	{28 20}	{26 18}

allocated to the left of the tree and even addresses to the right. Once the initial depth of the tree is set, allocating addresses is straight forward and does not need any serious computation.

For the network depth (2) the addresses are:

$$\left\{ \begin{array}{cc} (1) & \\ (3) & (2) \end{array} \right\} = \left\{ \begin{array}{cc} (A_k \ or \ B_k) & \\ (A_k + 2a_{k-1}) & (B_k + 2b_{k-1}) \end{array} \right\}$$

where the address of the parent node is in the first row and the addresses of the children nodes are in the second row. For the tree network shown up to depth(4) the assigned addresses are in Table 12.3.

Having allocated the first tier addresses which define the depth of the network up to d, we know that the remaining addresses are $2^{16} - 2^d$. These may be allocated by repeating the algorithm at each node or statically with each node receiving an equal set of $\left[(2^{16} - 2^d)/2^d \right]$ addresses. The set may be contiguous or distributed also using an arithmetic series so that for example at node 1 when d = 4 the following is true. For d = 4, 16 addresses have been allocated in tier 1. Node 1 then picks every 16th address for its set and this repeats down the tree through nodes 2, 3 until 15. The addressing plan therefore is as shown in Table 12.4.

Provided the 16 bit block address space is divisible by the number of nodes in the tier1 network, this plan does not leave unallocated addresses. However, it is possible to allocate addresses to nodes which are unable to use the whole set of addresses. Addresses allocated in the plan to end devices, should be redistributed to other nodes or kept in a pool by the Coordinator for nodes that run out of addresses for new access connection requests.

Table 12.4 Address Planning in Zigbee Netorks.

Node	Addresses allocated to each Node														
0	17	33	49	65	81	97	113	129	145	161	177	193	.	.	.
1	18	34	50	66	82	98	114	130	146	162	178	194	.	.	.
2	19	35	51	67	83	99		131		163		195	.	.	.
3	20	36	52	68	84	100						196	.	.	.
4	21	37	53	69	85	101							.	.	.
5	22	38	54	70	86	102							.	.	.
6	23	39	55	71	87	103							.	.	.
7	24	40	56	72	88	104							.	.	.
8	25	41	57	73	89	105	121						.	.	.
9	26	42	58	74	90								.	.	.
10	27	43	59	75	91								.	.	.
11	28	44	60	76	92								.	.	.
12	29	45	61	77	93								.	.	.
13	30	46	62	78	94								.	.	.
14	31	47	63	79	95								.	.	.
15	32	48	64	80	96	112							.	.	.

12.9 Iterative Address Allocation

In the iterative address allocation scheme, we reuse the algorithm given in equations (18) and (19). Assume the last address assigned to a node in tier1 is A_L. The algorithm which assigns the remaining addresses to the pre-deployed nodes in tier1 is given by the following:

$$a_{-1} = b_1 = 0; \ A_0 = B_0 = 0$$
$$A_L = last \ address \ allocated$$
Start: Node n
Depth: $t = d'$
Set $N_a = A_L + 1$
$A_{left} = N_a + 2a_{k-1}; \quad a_1 = 2$
$A_{right} = N_a + 2b_{k-1}; \quad b_1 = 1$
$if \ depth \ d' < d;$
    $continue$
$if \ (depth \ d' = d)$
    $\{n = n + 1;$
    $//Update \ last \ address$
$A_L = last \ address \ allocated$
    $Goto \ Set\}$
$Else$
    end

(12.20)

The advantage of this scheme is that all the addresses can be allocated uniquely and to known topological regions. The allocation could also be made based on requests from nodes. While this may incure some overhead if the last address allocated is always obtained from a central point, it is essential because the network is aware of how and where the address was allocated. The address allocation scheme is repeatable into the network nodes. Address conflicts can be avoided provided the last address is always updated into a known location that is accessed by all nodes. That is a all the nodes in the network have access to the "last address" allocated memory kept by the network and a node that accesses it takes the current reasonable addresses and then updates the site, then relinquishes its access right to the "last address" space. The disadvantage of this scheme is that each node vies for access to this space as in first come first served. This could pose some delays for nodes if they want addresses urgently and there are other nodes ahead of them in the queue waiting to access the same space before them.

12.10 Planning of Zigbee Personal Area Networks

There have been studies on several addressing schemes aimed at identifying nodes in Zigbee personal area networks or Zigbee sensor networks. So far their focus have been on finding suitable algorithms for addressing the nodes. The studies on node addressing have conspicuously omitted how to find out based on available network data the numbers of routers, network depth and children nodes. Most authors have assumed values for these quantities without relating those assumptions to realistic plans that can be used for detailed planning of personal area networks. In this section we propose new ideas which bridge these missing but significant details in current literature on PAN.

The short addressing space in PAN is normally reserved for identifying the network coordinator, routers and children. This address space limits the size of the network and also the number of children and routers that a network can support. Consider that it is necessary to use all the 16 bit short address space in a network. In practice this assumption will not be true always as some networks are going to be fairly small requiring only a sub-set of the address space. Assume that each RFD has circular or hexagonal coverage area (A_D) that is known based on the range of the Zigbee node. Assume also that the total region that need to be covered with the Zigbee PAN is known to be A_P. Define the cluster area to be formed by the coverage areas of the RFDs and that they are all able to hear the communication from the router they are attached to. Hence

$$N_T = \frac{Total\ coverage\_area\ of\ network}{Cluster\_area} \quad (12.21)$$

$$= \frac{A_T}{A_C};$$

$$\frac{A_C}{A_D} = \frac{(2^{16} - 2^d)}{2^d} = C_{Mc} \quad (12.22)$$

C_{Mc} is the total number of children nodes per cluster. Hence the total number of routers required for the network can be estimated to be

$$R_M = \frac{A_T}{A_C} \quad (12.23)$$

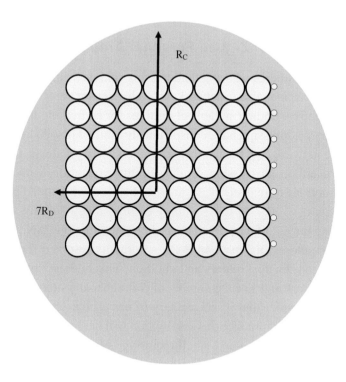

Fig. 12.11 End to End Zigbee Network formed by Children Nodes.

The cluster area is obtained from the coverage range of the head of the cluster (router) and the device area is also obtained from the coverage range of the end devices. Hence the number of children nodes required to cover the cluster region without intersecting based on omni-directional coverage is:

$$C_{Mc} = \frac{R_C^2}{R_D^2} \qquad (12.24)$$

Where R_C is the range of the cluster head and R_D is the range of the children devices.

Normally, the coverage range of nodes will intersect and in some circumstances, the radiation pattern of each Zigbee antenna is not omni-directional, but directional. Hence Figure 12.11 is modified for such applications.

12.11 Conclusions

This chapter has discussed planning of sensor networks including address planning scheme for personal area networks of Zigbee using tree based arithmetic series. The plans are systematic and not ad hoc and the addressing scheme eliminates address duplication and wastage in most addressing schemes. The scheme provides a means for estimating the number of nodes and also planning the network coverage.

References

[1] Y.-C. Wang, C.-C Hu and Y.-C. Tseng, "Efficient placement and dispatch of sensors in a wireless sensor network", IEEE Transactions on mobile computing, Vol. 7, No. 2, February 2008, pp. 262–274

[2] R. Peng, S. M.-Heng and Z. Y.-Min, "Zigbee Routing Selection Strategy Based on Data Services and Energy-balanced Zigbee Routing", in Proc. IEEE Asia-Pacific Conference on Services Computing (APSCC'06), December 2006, pp.

[3] D. Anurag, S. Roy and S. Bandyopadhyay, "Agro-sense: Precision agriculture using sensor-based wireless mesh networks", in Proc. 1st ITU-T Kaleidoscope Academic Conference.

[4] H.-I. Jeon, Y. Kim, "Efficient, Real-Time Short Address Allocations for USN Devices Using LAA (Last Address Assigned) Algorithm", in Proc. ICACT2007, February 12–14, 2007, pp. 689–683.

[5] J.I. Agbinya and M.A. Agbinya, "A New Address Allocation Scheme and Planning of Personal Area and Sensor Networks", in Proc. IEEE International Symposium on Parallel and Distributed Processing with Applications, ISPA '08, 10–12 December 2008, pp. 463–470.

13

Sensor Scheduling and Redeployment Mechanisms in Wireless Sensor Networks

Yinying Yang and Mihaela Cardei

Florida Atlantic University, USA

In this chapter, we study sensor scheduling and redeployment mechanisms in wireless sensor networks for both atomic event and composite event detection. Sensors may be equipped with different numbers and types of sensing components. They can detect an atomic event independently or they can cooperate to detect a composite event. Sensors can be put to sleep to save energy. In this chapter, we study scheduling mechanisms that select a set of active sensors to perform sensing and data relaying, while all other sensors go to sleep to save energy. Sensors in the active set change over time in order to prolong network lifetime. In addition, we study redeployment mechanisms which exploit sensor mobility to relocate sensors to improve the initial deployment.

13.1 Introduction

Sensors are used to monitor and control the physical environment. A Wireless Sensor Network (WSN) is composed of a large number of sensor nodes that are densely deployed either inside the phenomenon or very close to it [1]. Sensor nodes measure various parameters of the environment and transmit data collected to one or more sinks, using hop-by-hop communication. Once a sink receives sensed data, it processes and forwards it to the users. In mobile sensor networks, sensors can move via self-propelling, via means of wheels [5], springs [4], or they can be attached to transporters, such as robots [5] and vehicles [9].

Sensors may be equipped with different numbers and types of sensing components due to the following reasons: they might be manufactured with different sensing capabilities, a sensor node might be unable to use some of its sensing components due to the lack of memory for storing data, or some sensing components might fail over time. Sensors equipped with different sensing components can cooperate to detect a composite event [12].

Let us consider a single sensing component, for example, the temperature. If the sensed temperature value exceeds a predefined threshold, i.e., temperature > 100° C, we say that an *atomic* event has occurred. A *composite* event is the combination of several atomic events. For example, the composite event fire may be defined as the combination of temperature and smoke events. The composite event fire occurs only when both the temperature and the smoke rise above some predefined thresholds, such as temperature > 100° C and smoke > 100 mg/L, rather than a simple condition temperature > 100° C or smoke > 100 mg/L alone.

In this book chapter, we focus on sensor scheduling and redeployment mechanisms in WSNs for both atomic event detection and composite event detection.

An important issue in WSNs is energy management. Sensor nodes are battery powered and in general, they cannot be recharged. It will take a limited time before they deplete their energy and become un-functional. One of the major components that consume energy is the radio. A radio can be in one of the following modes: transmit, receive, idle, and sleep. A radio is in idle mode when the sensor is not transmitting or receiving data, and usually the power consumption is as high as in the receive mode. A radio is in sleep mode when both the transmitter and the receiver are turned off. According to [27], studies on several commercial radios (e.g. WaveLAN, Metricom) show that in sleep mode the power consumption ranged between 150–170 mW, while in idle mode the power consumption went up by an order of magnitude. One method to conserve energy is to put sensors to sleep mode when they are not actively participating in sensing or data relaying. We study sensor scheduling mechanisms that allow sensors to go to sleep to conserve energy and to prolong the network lifetime.

A large number of sensors can be distributed in mass by scattering them from airplanes or rockets. The initial deployment is hard to control in this case. However, a good deployment is vital in order to improve coverage, achieve load balance, and prolong network lifetime. Sensors can self-propel, they can move using wheels, springs, or they can be attached to transporters such as robots and vehicles. In this book chapter, we discuss algorithms that use sensors' mobility to relocate sensors after the initial deployment in order to improve the initial distribution.

13.2 Sensor Scheduling Mechanisms

Sensors are battery powered and in general, it is hard to recharge them. To prolong network lifetime, one method is to put "redundant" sensors to sleep and to activate them later, as needed. Sensor scheduling mechanism chooses a set of active sensors to perform sensing and data relaying, and all other sensors go to sleep to save energy. After some time, another set of active sensors is chosen, giving priority to the sensors with more energy resources. In this way, sensors work alternatively, resulting in balanced energy consumption. There are many mechanisms focusing on this problem [2, 3, 6, 7, 8, 10, 12, 13, 18, 21, 23]. This section elaborates some solutions in detail, including centralized approaches, localized algorithms, heuristic algorithms and distributed methods.

13.2.1 Sensor Scheduling Mechanisms for Atomic Event Detection

In [23], sensors conserve energy by identifying nodes that are equivalent from the routing point of view. The objective is to turn off unnecessary nodes while keeping a constant level of routing fidelity. *Routing fidelity* is defined as uninterrupted connectivity between communicating nodes. A Geographical Adaptive Fidelity (GAF) algorithm is proposed. Figure 13.1 shows an example for the communication between nodes 1 and 6. If node 5 is awake, then nodes 2, 3, and 4 can go to sleep to save energy.

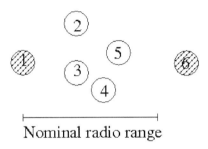

Fig. 13.1 Example of node redundancy in dense networks.

In this algorithm, it is assumed that the sensors know their location information using GPS or other localization protocols [24, 25, 26]. Each node can use location information to associate itself with a *virtual grid*, where all the nodes in the same grid are equivalent with respect to the task of forwarding messages. Nodes in the same grid use application and system information to coordinate with each other and to determine who will sleep and for how long. Nodes wake up periodically to decide whether they become active or if they continue to sleep.

Figure 13.2 shows an example of equivalent nodes. Nodes are equidistant and they are 1 unit apart. The radio range is slightly larger than 2 units. For the communication between nodes 1 and 4, nodes 2 and 3 are equivalent. For the communication between nodes 1 and 5, only the node 3 is acceptable.

To determine the node equivalence, the monitored area is divided into $r \times r$ virtual grids. The grid is defined so that for any two adjacent grids A and B, all the nodes in A can communicate with all the nodes in B and vice versa. Let Rc denote the radio range. To meet this definition, it is required that $r^2 + (2r)^2 \leq Rc^2$, that means $r \leq \frac{Rc}{\sqrt{5}}$. In this way, all nodes in the same grid are equivalent for routing. Figure 13.3 shows the virtual grid for Figure 13.1. There are three grids A, B, and C. According to the definition of the virtual grid, node 1 can reach any node in B, which contains nodes 2, 3, 4 and 5, and nodes 2, 3, 4 and 5 can all reach node 6. Therefore nodes 2, 3, 4 and 5 are equivalent for routing, and three of them can go to sleep.

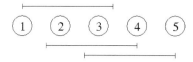

Fig. 13.2 The example of node equivalence.

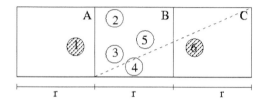

Fig. 13.3 An example of virtual grid.

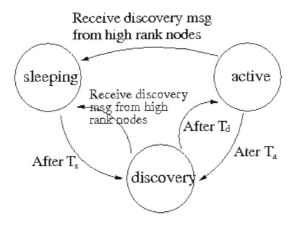

Fig. 13.4 State transition.

A node may be in one of the following three states: *sleeping*, *discovery*, and *active*. The state transition is shown in Figure 13.4. Initially, all the nodes are in the *discovery* state. They turn on the radio and exchange *discovery* (node id, grid id, *enat*-estimated node active time, node state) messages to find other nodes in the same grid. A node uses its location information and grid size to determine its grid id. A node entering the discovery state sets a timer T_d, which is a random value between 0 and some constant. T_d is used to reduce the probability of discovery message collision. When the timer T_d fires, a node broadcasts a discovery message and enters active state. A node entering active state sets a timer T_a, which shows how long the node stays in the active state. After T_a, the node returns to the discovery state. A node in the discovery or active state can go to the sleeping state when it finds out that some other equivalent nodes will handle routing. A node entering the sleeping state sets a timer T_s, which is an application dependent value. Nodes in sleeping state turn off the radio and wake up after T_s.

The T_a is set up to *enat* (estimated node active time). *enat* can be set up to be *entl*/2, where *entl* (expected node lifetime) is the node lifetime assuming the node will constantly consume energy at maximum rate until it dies. If *entl* is less than a predefined threshold (e.g. 30s), then set *enat* = *entl*.

The higher-ranked nodes have higher priority in becoming active, so that the lower-ranked nodes may go to sleep. Node ranking is used to maximize network lifetime by selecting which nodes handle routing. A node in the active state has higher rank than a node in the discovery state. For nodes in the same state, nodes with longer expected lifetime have higher rank. Node ids are used to break the ties. T_s is the sleep duration. One option is to set T_s to *enat* of the active node or T_s can be a random number in [*enat*/2, *enat*].

13.2.2 Sensor Scheduling Mechanisms for Composite Event Detection

Article [12] focuses on the composite event detection. To ensure a high quality of surveillance, some applications require that if an event occurs, it has to be detected by at least k sensors, where k is a user-defined parameter. A subset of sensors which jointly accomplish the event detection and alarming task is called a *detection set*. *Notification time* is defined as the summation of the time for all the members within a detection set to report the atomic event to the gateway, the time for the gateway to make a decision, and the time for a gateway to notify the base station that an event has occurred. An atomic event is said to be k-*watched* by a set of sensors D if at any time this event occurs at any point in the interest area, at least k sensors in the set D can detect this occurrence. A composite event is said to be k-*watched* by a set of sensors D if every atomic event forming that composite event is k-watched by set D. The problem addressed in [12] is defined as follows: given a set of sensors S, a monitored area A, and a composite event $\sum$ which is a combination of r atomic events $\sigma_l, l = 1, \ldots, r$, find a set of non-disjoint connected subsets (detection sets) $D_j, j = 1, \ldots, m$, of S and decide their corresponding active duration and subset masters (gateway nodes) such that:

- The composite event $\sum$ is k-watched by $D_j, j = 1, \ldots, m$, at any time.
- The network lifetime is maximized.
- For each detection set, the notification time is minimized.

Nodes have different sensing abilities, which mean that sensors may be equipped with different number and types of sensing components. For example, a sensor can sense light intensity and/or smoke density, while another sensor can sense temperature and/or pressure. Sensors with different sensing abilities can cooperate to detect composite events. For each type of sensing component, a sensor is equipped with no more than one sensing component.

For example, a sensor has only one temperature sensing component and/or one pressure sensing component. All of a sensor's sensing components turn on or off simultaneously. Sensors may have different communication ranges and different initial battery supplies.

The base station is in charge of constructing the detection sets. A greedy and centralized algorithm is proposed, which tries to form as many detection sets as possible. The algorithm starts by choosing a gateway node, which can be the node with the maximum residual energy. For each atomic event σ_l that forms the composite event, there is a counter c_l that keeps track of how many more sensors are needed in the current detection set in order to provide the k-watching of σ_l. The initial value for c_l is k for each σ_l. For a sensor s_i, a sensing component *component*$_{i,l}$ is called a *helpful* sensing component for the current c_l if it can monitor σ_l and the current value of c_l is greater than 0. Sensors are colored as follows:

- White: A sensor has not been considered yet.
- Black: A sensor has already been added to the detection set and is connected to the gateway through other nodes in the current detection set.
- Red: A useless sensor (the one whose all correlated counters c_l are 0) that has been considered, but has not been added to the detection set.
- Green: A sensor that has been added to the detection set but its parent is not a Black node, thus it is not connected to the gateway through other nodes in the current detection set.

A detection set is a breadth first search (BFS) tree rooted at the gateway. The steps for constructing detection sets are as follows.

1. m is the number of the detection set. Initially, $m = 0$.
2. While the sensor set S is not empty, do

    a. T is set to empty set, $T = \emptyset$. Set all counter c_l to k.

    b. Color all nodes White.

    c. Choose a node that is closer to the event consumer and has more energy as the gateway, gw, and color it Black.

    d. Add gw to set T, $T = \{gw\}$.

e. While at least one counter $c_l > 0$ do

    i. Invoke the Construct-leaves algorithm, which is explained later, and the result is stored in $\mathcal{L} = \text{Construct-Leaves}(S, T)$, which is a candidate list.

    ii. (Re)Calculate the contribution of each sensor in $\mathcal{L}$. The contribution of sensor i is computed as: $\chi_i = f(e_i, d_i) \times \frac{h_i}{sc_i}$, where $f(e_i, d_i)$ is a function to calculate s_i's lifetime depending on its current residual energy e_i and its current communication range d_i. h_i is the number of s_i's helpful sensing components. sc_i is the number of all sensing components that s_i is equipped with.

    iii. Color all the sensors with contribution 0 to Red and remove them from $\mathcal{L}$.

    iv. While $\mathcal{L}$ is empty and T's leaves and Red nodes still have White neighbors, do

        1. Invoke the Construct-leaves algorithm to build a new $\mathcal{L}$ using the breadth first search (BFS) algorithm.

        2. Calculate the contribution of each node in $\mathcal{L}$.

        3. Color all the nodes with contribution 0 to Red and remove them from $\mathcal{L}$.

    v. If $\mathcal{L}$ is still empty, then go to step 2.e.viii.

    vi. Sort $\mathcal{L}$ in descending order of contributions.

    vii. While $\mathcal{L}$ is not empty do

        Remove the sensor ρ from the top of $\mathcal{L}$ and add it to T. If

ρ's parent is Black, then color ρ Black. Otherwise, color ρ Green. Decrease all ρ's correlated counters by 1. If a counter is 0, then proceed as follows: if all counters are 0, then go to step 2.e.viii, else go to step 2.e.ii.

viii. If any counter > 0, then remove T from S and go to step 2.

ix. Else, if there is a Green sensor then

    1. For each Green sensor ρ, do

        a. ρ's parent is denoted as $\kappa = \rho.Parent$.

        b. While κ is Red, do

            i. Color κ Black and add κ to T.

            ii. $\kappa = \kappa$'s parent

x. $m = m + 1$

xi. A new detection set has been found. Its activation time, denoted t_m, is set up to the smallest lifetime of a sensor in this detection set. Recalculate the residual energy of each sensor in the detection set. Remove from S the sensors that ran out of energy.

3. Return the m detection sets.

The Construct-leaves algorithm returns $\mathscr{L}$ and has the following steps.

1. Construct a list $\mathscr{L}$ consisting of all the White neighbors of T's leaves and all the White neighbors of the Red nodes.
2. Assign a parent for each sensor ρ in $\mathscr{L}$ as follows. For each ρ in $\mathscr{L}$, if any neighbor of ρ is Black or Green, then ρ's parent is the Black/Green

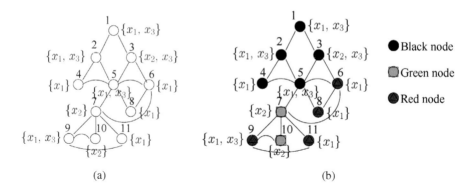

Fig. 13.5 The construction of a detection set. Node 1 is the gateway.

sensor with the least number of children. Else ρ's parent is the Red node with the longest lifetime.

Figure 13.5 shows an example. Figure 13.5a is the initial deployment. The composite event is the combination of $\{x_1, x_2, x_3\}$ and $k = 3$. The circles denote sensors. The number above each sensor is the sensor id and the numbers inside the braces are the sensing components it is equipped with. For example, sensor 1 is equipped with sensing components x_1 and x_3. Figure 13.5b shows the detection set consisting of Black and Green nodes after executing the above algorithm.

Article [18] considers that sensors can have different sensing ranges for different sensing components. For simplicity, assume that $Rs_1 \leq Rs_2 \leq \cdots \leq Rs_M$, where $Rs_1, Rs_2, \ldots, Rs_M$ are the sensing ranges for the M sensing components $x_1, x_2, \ldots, x_M$, respectively. The problem addressed in [18] is as follows: given a WSN deployed for watching a composite event $\{x_1, x_2, \ldots, x_M\}$, design a sensor scheduling mechanism that selects the set of active sensors such that to ensure the coverage and connectivity requirements for each sensing component and such that the WSN lifetime to be maximized. The *coverage* requirement requires that the deployed area to be continuously covered by each sensing component x_j, for $1 \leq j \leq M$. The *connectivity* requirement requires that the set of active sensors to be connected, which is necessary in order to collect the sensed data. If the sink is connected to any of the active sensors, then a data collection tree rooted at the sink can be formed and data can be gathered by the sink.

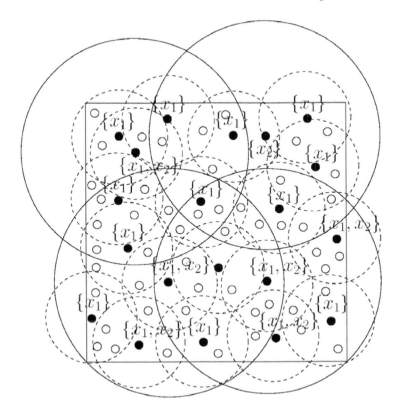

Fig. 13.6 The monitored area is covered considering both sensing components x_1 and x_2.

Figure 13.6 shows an example with $M = 2$ sensing components $\{x_1, x_2\}$. The sensor communication range is $Rc = 2Rs_1$, where Rs_1 is the sensing range of sensing component x_1. The black nodes are active sensors and the white nodes are sensors in sleep mode. The dashed circles represent the sensing area of sensing component x_1, and the continuous line circles represent the sensing area of x_2. The set of active sensors meets the coverage requirement: the whole monitored area is covered by the sensing component x_1 and the sensing component x_2. The connectivity requirement is met as well: the active sensors are connected so that the sensed data can be collected by a sink.

Two solutions are proposed for this problem, a distributed approach and a localized method. Network lifetime is organized in rounds. Each round has two phases: initialization and data collection. In the initialization phase, a scheduling algorithm is run to decide which sensors remain active and which sensors go to sleep during the current round. In the data collection phase, active sensor nodes perform sensing and data relaying. In the article, it is assumed that sensors know their location information using GPS or other localization protocols [24, 25, 26].

In the distributed approach, the square monitored area is divided into grids. It is assumed that each sensing component x_j has coverage range $r_j \sqrt{2}$, such that any sensor with sensing component x_j located in a $r_j \times r_j$ grid region completely covers that region. For simplicity, it is assumed that $r_j = 2^u \cdot r_{j-1}$, where $u \geq 0$. It is straight-forward to extend this approach to other cases. If $u = 0$, then the two consecutive sensing components x_j and x_{j-1} have the same sensing range. Figure 13.7 shows an example with $M = 3$ sensing components, $x_1, x_2,$ and x_3, where $r_3 = r_2 = 2r_1$. Figure 13.7 illustrates both active and sleep mode sensors in a quarter of the deployment area. The set of active sensors satisfies the coverage condition: each grid $r_1 \times r_1$ has an active sensor equipped with sensing component x_1, each grid region $r_2 \times r_2$ has an active sensor equipped with sensing component x_2, and each grid region $r_2 \times r_2$ has an active sensor equipped with sensing component x_3. The scheduling mechanism consists of two algorithms, one to decide the sensing nodes, and one to decide the relay nodes needed to ensure the connectivity.

In order to decide the sensing nodes, the initialization phase is divided into a sequence of time intervals $t_1, t_2, \ldots, t_M$. The basic idea is to ensure that during the time interval t_j at least one sensing component x_j becomes active in each square region $r_j \times r_j$. Each sensor s_i keeps a Boolean variable g_j for each

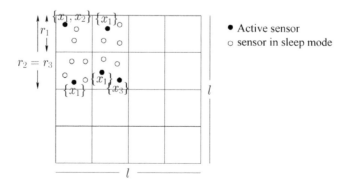

Fig. 13.7 Sensors deployed in the square area and grid division for the distributed approach.

sensing component x_j that it is equipped with. Initially, all variables $g_j = 0$. For example, a sensor $s_i\{x_1, x_3, x_4\}$ initializes variables $g_1 = g_3 = g_4 = 0$. Variable g_j becomes 1 if there is at least one active component x_j in the $r_j \times r_j$ region where s_i is located. This happens when s_i or another sensor with sensing component x_j, located in the same $r_j \times r_j$ region, becomes active.

The following steps are run by each sensor s_i at the beginning of the time interval t_j. Based on the partitioning of the field, a sensor s_i belongs to exactly one $r_j \times r_j$ square region. In time interval t_j, at least one sensing component x_j becomes active in each $r_j \times r_j$ region, if such a component exists.

1. If s_i has the status active, or s_i is not equipped with x_j, or if another x_j component has become active ($g_j = 1$), then s_i does not change its status and the procedure ends.

2. Otherwise, sensor s_i computes its contribution function $\chi_i = f$ $(e_i, Rc) \cdot help_i$, where $f(e_i, Rc)$ is a function to calculate s_i's lifetime based on its current residual energy e_i and its current communication range Rc, $help_i$ is the number of s_i's helpful sensing components. A sensing component x_k of s_i is called a helpful sensing component if $k \geq j$ and $g_k = 0$. A larger contribution function value means s_i has a larger priority in becoming active. Then s_i starts a timer inversely proportional to χ_i.

3. If a sensor s_i receives a *StatusActive*(s_k, s_k's location, s_k's sensing components $\geq j$) message from a sensor s_k before its timer fires up, then s_k has become active recently. Then for each s_k's sensing component $v \geq j$, if s_i is equipped with x_v, $g_v = 0$, and s_i and s_k are in the same $r_v \times r_v$ region, then s_i sets $g_v = 1$. If g_j becomes 1, then the procedure ends without requiring s_i to become active.

4. If the timer fires up, then sensor s_i is set active and thus g_j becomes 1. When a sensor becomes active, it broadcasts a *StatusActive*(s_i, s_i's location, s_i's sensing components $\geq j$) message in s_i's region $r_w \times r_w$, where w is the largest index of the sensing components of s_i. A sensor s_w forwards the message only the first time it receives the message and only if it is located in the same $r_w \times r_w$ region. Every forwarding sensor s_w sets up its g variables based on s_i's sensing components as follows. For each s_i's sensing component $v \geq j$, if s_w is equipped with x_v and $g_v = 0$, then s_w sets its $g_v = 1$.

When the sensor communication range $Rc \geq r_1\sqrt{5}$, the coverage requirement implies the connectivity requirement. According to Figure 13.8a, the communication range (Rc) needed for direct communication of two sensors located at the maximum distance in two adjacent grids is $Rc^2 = r_1^2 + 4r_1^2 = 5r_1^2$. Therefore, a communication range $Rc = r_1\sqrt{5}$ ensures direct communication of any two sensors located in adjacent regions.

Connectivity becomes an issue when $Rc < r_1\sqrt{5}$. Two active sensors in adjacent regions might not communicate directly. We consider that sensors in the same region $r_1 \times r_1$ can communicate directly, that is $Rc \geq r_1\sqrt{2}$. Length r_1 can always be chosen such that this condition is met. For connectivity purpose, it is sufficient to ensure that each square region $r_1 \times r_1$ is connected to its bottom and right neighbors. After satisfying the sensing requirement, there will be at least one active sensor in each $r_1 \times r_1$ square region. To ensure that active sensors form a connected topology, the basic idea is to ensure that each $r_1 \times r_1$ square region is connected to its bottom and right neighbor regions, see Figure 13.8b. The main steps are as follows.

1. Each $r_1 \times r_1$ region chooses a cluster head (CH), which is for example, the active sensing node with the largest remaining energy. The CH of each region is responsible to ensure connectivity with bottom and right regions. Each sensor sends a *Hello*(s_i, s_i's region location, s_i's residual energy) message. If a CH s_i hears *Hello* messages from any nodes in the bottom and right regions, then it does nothing.

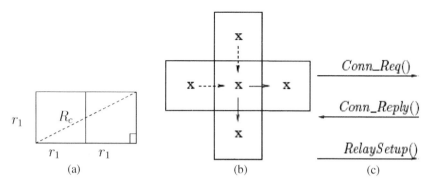

Fig. 13.8 Connectivity mechanism. (a) Maximum distance between sensors in adjacent regions. (b) Connection of a region with bottom and right regions. (c) Connectivity: three-way message exchange mechanism.

2. Otherwise, CH s_i starts a three-way message exchange mechanism, see Figure 13.8c: request using *Conn_Req*, reply using *Conn_Reply*, and relay setup using *RelaySetup* message. Node s_i broadcasts *Conn_Req* (CH s_i, grid position of s_i, list of needed region connections (bottom and/or right), sending node ID, TTL, number of relay hops). The Time-To-Live (TTL) is used to control the number of hops the message can be forwarded.

3. An intermediate node s_j receiving a *Conn_Req* message will forward the message if s_j is not an active sensing node in the destination bottom and/or right regions; otherwise it will reply with a *Conn_Reply* message. If s_j is not an active sensing node in a destination region, then it forwards the *Conn_Req* if this is the message with the minimum number of relay hops so far and increases the number of relay hops. In addition, s_j updates the following fields: sending node ID (which is now s_j), TTL = TTL - 1, and the number of relay hops. The message will be forwarded only if TTL > 0. If s_j is an active sensing node in one of the destination regions (in the bottom or right region), then s_j will not forward the *Conn_Req* message, but it will reply with a message *Conn_Reply* (s_j, grid position of s_j, CH s_i, grid position of s_i, number of relay hops). The reply will be sent to s_i along the reverse pointers which were set-up temporarily when the requests were forwarded.

4. If the CH node s_i receives one or more *Conn_Reply* messages from a region of interest (bottom or right), then it chooses the one with the minimum number of relay hops. Then s_i sends a message *RelaySetup* (CH s_i, grid position of s_i, s_j, grid position of s_j). This message is sent along the forward pointers which were set-up temporarily when the replies were forwarded. Each node that forwards the *RelaySetup* message becomes active and serves as a relay node in order to ensure connectivity.

5. If no reply message is received from a destination region (bottom or right), then according to the incremental ring search approach, TTL is increased and the process is repeated. TTL can be increased up to some predefined maximum value. If no reply is received after a number of trials, then connectivity cannot be satisfied. One approach

in this case is to have s_i send a message to the sink announcing this event. If the sink location is not known, this can be accomplished by flooding the message in the whole network.

The second method to decide the set of active sensors is a localized approach. The initialization phase is divided into M time intervals $t_1, t_2, \ldots, t_M$. In each time interval $t_j (1 \leq j \leq M)$, sensors achieve the coverage and connectivity of the sensing component x_j. A sensor is in one of three possible states: active, pending, and sleep. Active and sleep are final states. Pending is an intermediate state. Initially, all the sensors are in the pending state. Sensors which are chosen to be sensing nodes or relay nodes switch to the active state. Sensors which do not become active in the current time interval remain in the pending state and they might become active in the following time intervals. Sensors in the sleep state go to sleep to save energy.

Every sensor s_i has a priority P_i^j in each time interval t_j, which is a 4-tuple $P_i^j = (state, f(e_i, Rc), help_i, ID_i)$. In P_i^j, *state* is the most important factor. Sensors in the active state have higher priorities than those in the pending or sleep state. *state* = 1 when the sensor is in the active state and *state* = 0 when it is in the pending or sleep state. $f(e_i, Rc)$ is the same as that defined in the previous distributed algorithm, and it is used to break the tie for the sensors with the same *state*. Sensors having higher $f(e_i, Rc)$ have higher probabilities to be active in the following time intervals. $help_i$ is the number of s_i's helpful sensing components. A sensing component x_k in time interval t_j is helpful when $k \geq j$. $help_i$ is used to break the tie for sensors with the same *state* and the same $f(e_i, Rc)$. ID_i is the node identification. All sensors are initialized to be in the pending state.

Each sensor exchanges *Hello* messages with its neighbors. A *Hello* message includes the state of the sensor, the list of the sensing components the sensor has, and the priority of the sensor. The neighbors are defined as follows: $N_j(s_i) = \{s_k \in \aleph | d(s_i, s_k) \leq Rs_j, s_k \neq s_i\}$. $\aleph$ is the sensor set containing sensors deployed in the whole monitored area. $d(s_i, s_k)$ is the distance between two sensors and Rs_j is the sensing range in the current time interval t_j, which is the sensing range of the sensing component x_j. The neighbors may be more than one hop away depending on the Rs_j and Rc. When sensors within Rs_j receive the Hello message, they broadcast it. If non-neighbor sensors receive the Hello message, they discard it and do not broadcast it. In the time interval t_j, the sensors that were in the active or sleep state in the previous time interval t_{j-1} remain in the same state in t_j. Each sensor s_i that is in the pending state and is equipped with the sensing component x_j considers the following three conditions. If all three conditions are met, it goes to the sleep state. Otherwise, if *Condition*$_1$ and *Condition*$_2$ are met, but *Condition*$_3$ is not met, it remains in the pending state. If at least one of the *Condition*$_1$ and *Condition*$_2$ is not met, then the sensor s_i goes to the active state.

- *Condition*$_1$: s_i's sensing area can be covered by its neighbors in active or pending state, containing the sensing component x_j, and having higher priorities.
- *Condition*$_2$: for any pair of s_i's neighbors that are in the active or pending state, there exists an h-hop ($h \geq 1$) path connecting them. The path should only contain sensors that are not in the sleep state and have higher priorities than s_i, which implies that s_i cannot be in the path.
- *Condition*$_3$: The sensor does not contain any sensing components $x_k, k > j$, that is, x_{j+1}, or x_{j+2}, ..., or x_M.

Condition$_1$ regards the coverage requirement and ensures that only sensors whose sensing area can be covered by their neighbors can go to sleep. *Condition*$_2$ regards the connectivity requirement. If sensor s_i's neighbors can be connected without s_i's help, then s_i has a chance to go into the sleep state. In *Condition*$_3$, if s_i has a sensing component whose index is larger than j, then its other sensing components have to be checked in the following time intervals. Note that since $Rs_{j+1} \geq Rs_j$, the algorithm only adds active sensors in each of the following time intervals, such that to meet the coverage and connectivity requirements for larger sensing ranges. Therefore, an active sensor cannot move to the pending or sleep state. If a sensor is in the pending state at the end of the time interval t_j, then in the time interval t_{j+1} it updates its priority by recalculating $help_i$. If a sensor goes to the active state at the end of the current time interval t_j, then it updates its priority by setting $state = 1$. A similar process repeats in the next time interval, using the updated priorities. At the end of the M time intervals, each sensor either goes to the sleep state or to the active state. Thus, the set of active sensors is computed and the sensing and data collection phase begins.

13.3   Sensor Redeployment Mechanisms

A large number of sensors can be distributed in mass by scattering them from airplanes, rockets, or missiles [1]. In such a case, the initial deployment is difficult to control. However, good deployment is necessary to improve coverage, achieve load balance, and prolong network lifetime. Recent research has focused on methods to improve the initial deployment. One possible method ([4, 14, 15, 17] and [22]) is to use mobile sensors, thus allowing sensors to relocate after the initial deployment.

13.3.1   Sensor Redeployment Mechanisms for Atomic Event Detection

There are recent research works proposing sensor redeployment mechanisms for atomic event detection. In [4], Chellappan *et al.* study the flip-based deployment mechanism to achieve a maximum coverage. They assume a sensor can only flip once from its current location to a new location, when triggered by an appropriate signal, and the maximum flipping distance is F. Sensors are deployed in a square field, divided into $R \times R$ square regions (see Figure 13.9). A sensor can flip from its own region to one of its neighbor regions, which are its left, right, top, and bottom neighbor regions. Although each sensor can flip at most once, the flip distance can be adjusted depending on the triggering signals. In [4], it is assumed that F is an integral multiple of the basic unit d. C is used to denote the flip choices. Thus, $C = n$ means that a sensor can flip once to one of the distance $d, 2d, 3d, \ldots$, or nd. In the initial deployment, there may be regions which are not covered by any sensor. The goal is to determine an optimal sensor movement plan such that to maximize the number of regions covered by at least one sensor, while simultaneously minimizing the total number of flips used.

Figure 13.9 shows an example. Figure 13.9a shows the initial deployment. Circles denote the sensors and the number in each region is the region id. Let the region size $R = d$ and the flip distance $F = d$. There are regions in the initial deployment that are not covered by sensors, such as regions 2, 11, 12, and 16. Figure 13.9b shows the optimal movement plan. Since sensors can flip only once, one sensor in region 4 flips to region 8 first and then one sensor in region 8 flips to region 12 in order to cover the region 12. Figure 13.9c shows the resulting deployment, where all regions are covered.

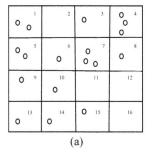 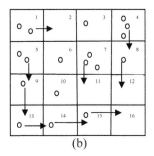

Fig. 13.9 An example for the flip-based movement plan. (a) The original deployment. (b) The optimal movement plan. (c) The resulting deployment.

The solution proposed is to model the network as a graph structure, called *virtual graph*, and then to apply a minimum-cost maximum-flow algorithm to solve the problem. Regions with at least two sensors are called *sources*, regions with only one sensor are called *forwarders*, and regions without any sensors are called *holes*.

The virtual graph for the case when $R = d$, $F = d$, and $C = 1$ is constructed as follows. Each region i is represented by three vertices v_i^b, v_i^{in} and v_i^{out}. v_i^b is the base vertex, which keeps track of the number of sensors in region i. v_i^{in} and v_i^{out} are used to keep track of the number of sensors from other regions that have flipped to region i and the number of sensors that have flipped from region i to other regions, respectively. For a *source* or *forwarder* region i, edges are added inside the region as follows:

- An edge from v_i^b to v_i^{in} with capacity $n_i - 1$, where n_i is the original number of sensors in region i. The insight of this is that when attempting to determine the flow from v_i^b, at least one sensor will remain in region i.
- An edge from v_i^{in} to v_i^{out} with capacity n_i. This means that it is possible for up to n_i sensors to flip out of this region.

Edges are added inside a *hole* region as follows:

- An edge from v_i^{in} to v_i^b with capacity 1. This will allow at most one sensor into v_i^b.
- An edge from v_i^{in} to v_i^{out} with capacity 0. This is because a sensor that moves into a hole region is not able to flip further.

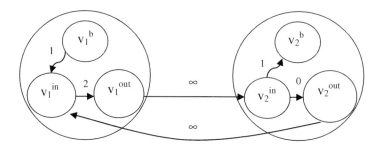

Fig. 13.10 The virtual graph of the regions 1 and 2.

Other edges are added between reachable regions. The *reachable regions* of a region i are those regions where sensors from region i could flip to, and they depend on the flip distance F. If region i and region j are reachable, then:

- An edge of infinite capacity is added from v_i^{out} to v_j^{in}. This allows any number of flips between reachable regions.
- An edge of infinite capacity is added from v_j^{out} to v_i^{in}.

Figure 13.10 shows an example of the virtual graph of the regions 1 and 2 from Figure 13.9a.

When a flip from one region to another region occurs, a cost of 1 has incurred. In order to compute the number of flips between two regions, a cost value is added to the corresponding edge. We have $Cost(v_i^{out}, v_j^{in}) = Cost(v_j^{out}, v_i^{in}) = 1$. The cost for other edges is 0, since a flip inside one region is not counted toward sensor flips.

After the virtual graph has been constructed, determining the minimum cost maximum flow plan in the graph is a two-step process. First, the maximum flow value between sources and holes is determined using the Edmonds-Karp algorithm [19]. Then the method in [20] is applied to the maximum flow in order to determine the minimum cost flow plan.

The above method can be extended to the cases (1) $R = d$, $F > d$, and $C = 1$, (2) $R = d$, $F > d$ and $C = n$, and (3) $R > d$ [4].

In [15], Wu and Yang present SMART, a scan-based distributed protocol that relocates sensors such that to achieve the load balance. The goal is to get a uniform distribution of the sensors via sensor relocation. The monitored network is divided into square regions, see Figure 13.11. Each region is covered by a cluster of sensors, controlled by a cluster head. Each cluster head knows

1	3	2	2
25	5	31	11
11	3	15	7
16	21	3	4

(a)

2	2	2	2
18	18	18	18
9	9	9	9
11	11	11	11

(b)

10	10	10	10
10	10	10	10
10	10	10	10
10	10	10	10

(c)

Fig. 13.11 An example for the SMART mechanism. (a) Initial deployment. (b) The deployment after the first round. (c) The deployment after the second round.

the location and the number of sensors in the cluster and it is in charge of the communication with adjacent clusters. Figure 13.11a shows the initial deployment. The number in each region shows the initial number of sensors in the region. The basic idea is as follows. Two rounds are used, first on the rows and the second one on the columns. As shown in Figure 13.11b, after the first round, all rows are balanced. After the second round, all columns and the whole network are balanced as shown in Figure 13.11c.

Let us discuss the scan mechanism in detail. Let w_i denote the number of sensors in cluster i. v_i denotes the prefix sum of the first i clusters in the same row or column, i.e., $v_i = \sum_{j=1}^{i} w_j$. $v_n = \sum_{j=1}^{n} w_j$ is the total sum. $\bar{w} = v_n/n$ is the average number of sensors in a balanced state and $\bar{v} = i\bar{w}$ is the prefix sum in the balance state.

Each round includes two scans. For clusters in the same row, the first scan works from one end of the row to the other end, in the *positive* direction, and the second scan works from the other end back to the initial end, in the *negative* direction. The first scan computes the total number of sensors in the clusters. Each cluster head i determines its prefix sum v_i by adding $v_{i-1} + w_i$ and forwarding v_i to the next cluster. The cluster head in the last cluster computes v_n and $\bar{w} = v_n/n$ and initiates the second scan by sending out $\bar{w}$. The second scan forwards the average number of sensors in a cluster to each cluster in the row. Each cluster head computes $\bar{v}_i = i\bar{w}$ based on $\bar{w}$ that is passed around and its own cluster position.

Each cluster then determines the give/take sensors to/from other clusters independently. When $w_i - \bar{w} = 0$, cluster i is in the "neutral" state. When $w_i - \bar{w} > 0$, it is in the "give" state and when $w_i - \bar{w} < 0$, it is in "take" state. A cluster in the give state decides the number of sensors to be sent to the positive direction ($w_i^{\rightarrow}$) and to the negative direction ($^{\leftarrow}w_i$) according to the following formulas.

$$w_i^{\rightarrow} = \min\{w_i - \bar{w}, \max\{v_i - \bar{v}_i, 0\}\} \tag{13.1}$$

$$^{\leftarrow}w_i = (w_i - \bar{w}) - w_i^{\rightarrow} \tag{13.2}$$

A cluster in the take state determines the number of sensors to be taken from the positive direction ($w_i^{\leftarrow}$) and from the negative direction ($^{\rightarrow}w_i$) according to the following formulas.

$$^{\rightarrow}w_i = \min(\bar{w} - w_i, \max\{v_{i-1} - \bar{v}_{i-1}, 0\}) \tag{13.3}$$

$$w_i^{\leftarrow} = (\bar{w} - w_i) - {}^{\rightarrow}w_i \tag{13.4}$$

After each cluster has computed the number of sensors to be sent or taken, the sensors move to achieve the load balance. The procedure is repeated for each row and then for each column. Let us consider Figure 13.11a as an example. In the third row, the clusters are labeled 1, 2, 3, and 4 from left to right. After the first and second scan, the prefix sum is computed by each cluster, as shown in Table 13.1, and $\bar{w} = 9$. Clusters 1 and 3 are in the give state. For cluster 1, according to formulas (13.1) (13.2), $w_1^{\rightarrow} = 2$, $^{\leftarrow}w_1 = 0$. For cluster 3, $w_3^{\rightarrow} = 2$, $^{\leftarrow}w_3 = 4$. Clusters 2 and 4 are in the take state. For cluster 2, according to formulas (13.3) (13.4), $^{\rightarrow}w_2 = 2$, $w_2^{\leftarrow} = 4$. For cluster 4, $^{\rightarrow}w_4 = 2$ and $w_4^{\leftarrow} = 0$. Then, cluster 1 sends 2 sensors to its right clusters, cluster 2 takes 2 sensors from its left clusters and 4 from its right clusters, cluster 3 sends 4 sensors to its left clusters and 2 to its right clusters, and cluster 4 takes 2 sensors from its left clusters.

If there are holes in the network, then there is no sensor that can be the cluster head in those regions and certain rows and columns may be disconnected. Therefore, before applying the scan based mechanism, a pre-processing scheme is needed to plant "seeds" to those empty regions. The clusters with

Table 13.1 The scan process on the third row of Figure 13.11a.

i	1	2	3	4
w_i	11	3	15	7
v_i	11	14	29	36
$\bar{v}_i$	9	18	27	36

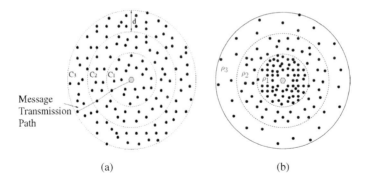

Fig. 13.12 WSN model using coronas concentric to the sink. (a) Uniform distribution. (b) Non-uniform distribution.

more than one sensor first send sensors to the holes to make sure that there is at least one sensor in each cluster and then the row-wise and column-wise scans are used to achieve a uniform distribution.

Articles [17] and [22] discuss a partitioning of the monitoring area in coronas, and consider repositioning sensors to prolong network lifetime. In a WSN, sensors closer to the sink tend to consume more energy than those farther away from the sink [11]. Article [11] considers a uniform sensor deployment and divides the monitored area in coronas, $C_1, C_2, \ldots, C_n$, as shown in Figure 13.12a. The sink is located at the center of the monitored area. A message transmitted from corona C_i is forwarded by sensor nodes in coronas C_{i-1}, C_{i-2}, and so on until it reaches corona C_1 from where it is transmitted to the sink. Corona width is chosen such that a message is forwarded by only one sensor in each corona. Assuming that each sensor is equally likely to be the source of a path to the sink, [11] shows that in order to minimize the total energy consumption, all the coronas must have the same width. Let d be the width of each corona. In this case sensors suffer an uneven energy depletion, with sensors in the first corona C_1 being the first to die. This is mainly because, besides transmitting their own packets, they will also forward packets on behalf of other sensors that are located farther away. A uniform sensor deployment (uniform density) results in holes in the WSN that will reduce network lifetime. To prolong network lifetime, [17] and [22] first compute the desired non-uniform sensor density in the monitored area in order to reduce the energy holes near the sink. Then sensor relocation algorithms are proposed that reposition sensors according to the desired density.

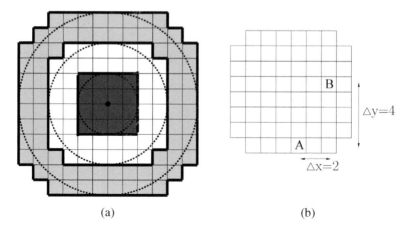

(a)   (b)

Fig. 13.13 (a) Division of monitored area into square regions. (b) Manhattan distance between regions A and B.

The basic idea in computing the desired non-uniform density for each corona is to make sure that all the sensors deplete their energy at the same rate. The monitored area is divided into a grid of $r \times r$ regions, as shown in Figure 13.13a. In this case, the division in coronas is not circular, but it follows the regions' contour. If the region granularity is very small, $r \to 0$, then the division in coronas is similar to the one in Figure 13.12a, where coronas are circular. It is assumed that the energy consumption is proportional to the number of messages transmitted and that sensors are uniformly deployed in the same corona. Intuitively, to balance the energy consumption, the corona C_n should have the lowest density and the corona C_1 should have the highest density. Let ρ_i denote the density of the corona C_i, thus $\rho_1 \geq \rho_2 \geq \cdots \geq \rho_n$, as shown in Figure 13.12b.

To balance the energy consumption, each sensor transmits the same number of messages in a specific time interval T. This is a periodic data reporting process, where one data message is generated from each unit of area during each time T. Based on this assumption, a sensor in corona C_i will generate $1/\rho_i$ messages in each time interval T. Besides, it also participates in forwarding messages generated in coronas $C_{i+1}, C_{i+2}, \ldots, C_n$. Note that sensors in corona C_n do not forward messages on behalf of other sensors. N_i denotes the number of sensors in corona C_i, A_i denotes the area of C_i, and A denotes the total area of the monitored area. The number of messages generated in C_i is N_i/ρ_i. The number of messages forwarded in C_i in each time interval is:

$$\frac{\frac{N_{i+1}}{\rho_{i+1}} + \frac{N_{i+2}}{\rho_{i+2}} + \cdots + \frac{N_n}{\rho_n}}{N_i} = \frac{A_{i+1} + A_{i+2} + \cdots + A_n}{A_i \cdot \rho_i}$$

$$= \frac{A - (A_1 + A_2 + \cdots + A_{i-1})}{A_i \cdot \rho_i} - \frac{1}{\rho_i}$$

The total number of messages transmitted by a sensor in C_i is:

$$TotalNum_i = \frac{A - (A_1 + A_2 + \cdots + A_{i-1})}{A_i \cdot \rho_i} - \frac{1}{\rho_i} + \frac{1}{\rho_i}$$

$$= \frac{A - (A_1 + A_2 + \cdots + A_{i-1})}{A_i \cdot \rho_i}$$

To balance the energy consumption, $TotalNum_i = TotalNum_n$, therefore:

$$\frac{A - (A_1 + A_2 + \cdots + A_{i-1})}{A_i \cdot \rho_i} = \frac{1}{\rho_n}$$

$$\Rightarrow \rho_i = \rho_n \cdot \frac{A - (A_1 + A_2 + \cdots + A_{i-1})}{A_i}$$

ρ_n is computed according to the following formula:

$$\rho_1 \cdot A_1 + \rho_2 \cdot A_2 + \cdots + \rho_n \cdot A_n = N$$

$$\Rightarrow \rho_n \frac{A}{A_1} A_1 + \rho_n \frac{A - A_1}{A_2} A_2 + \cdots + \rho_n \frac{A - A_1 - A_2 - \cdots A_{n-1}}{A_n} A_n = N$$

$$\Rightarrow \rho_n = \frac{N}{\sum_{i=1}^{n} i \cdot A_i}$$

In [22], several methods are proposed for the following problem: given a WSN with sensors randomly deployed for periodical monitoring of an area centered to a sink, determine a sensor movement plan that achieves the desired non-uniform sensor density, as discussed above, while minimizing the total sensor movement.

The Integer Programming (IP) approach considers the division from Figure 13.13a. Each region chooses a representative responsible for the communication with all the sensors in the region and with the sink. The desired number of sensors of a region in corona C_i is denoted as $N_i^r = \rho_i \cdot r^2$. A region in C_i can be a source, hole, or neutral region depending on whether the current number of sensors is greater than, less than, or equal to N_i^r.

A bipartite graph $G = (V, U, E)$ is constructed where V, U are two node sets and E is the edge set. Source regions (hole regions) are represented as nodes in the set V (set U). Each node v has associated a weight $w(v)$ corresponding to the amount of sensor overload (if $v \in V$) or sensor underload (if $v \in U$). An edge is added between any two nodes in V and U. The weight of an edge is defined as the Manhattan distance between the corresponding source and hole. For example, in Figure 13.13b, the Manhattan distance between the regions A and B is $\Delta x + \Delta y = 2 + 4 = 6$. The problem reduces to matching all under loaded regions such that the sum of the weights of the selected edges is minimized.

The variable x_{ij}, where $i = 1, \ldots, |U|$, $j = 1, \ldots, |V|$, and $x_{ij} \in \{0, 1, \ldots, \min\{w(v_i), w(u_j)\}\}$ denotes the number of sensors that will move from the source region v_i to the hole region u_j. c_{ij} is the weight of the edge (v_i, u_j). The optimal solution is computed using the following IP-formulation:

$$\text{Minimize} \quad \sum_{ij} c_{ij} x_{ij}$$

$$\text{Subject to} \quad \sum_{j=1}^{|V|} x_{ij} \leq w(v_i) \quad \text{for all } i = 1, \ldots, |U|$$

$$\sum_{i=1}^{|U|} x_{ij} \leq w(u_j) \quad \text{for all } j = 1, \ldots, |V|$$

Remarks:

- The objective function asks to minimize the total sensor moving distance.
- The first constraint requires that the number of sensors that leave the source region v_i be upper-bounded by $w(v_i)$, which is the overload of that region.
- The second constraint requires that the number of sensors that enter a hole region u_j be $w(u_j)$, which is the underload of that region.

The sink uses an IP-solver to compute the sensor movement plan (given by the x_{ij} values) and forwards it to the region representatives which coordinate the senor movement inside that region.

Another approach is the localized matching method [22]. Each region selects a representative in charge of communication with the neighbor regions' representatives and with organizing the movement inside the region. Sensors in a region can move only to neighbor regions: left, right, top, and bottom. The movement distance between two regions is computed using the Manhattan distance. Δ_+ is the number of overloaded sensors in a source region and Δ_- is the number of underloaded sensors in a hole region.

The movement protocol is initiated by the holes and is a three-way message exchange process. Assuming that the total number of overloaded sensors is greater than or equal to the number of underloaded sensors, all hole regions requirements will be satisfied when the algorithm completes. A hole region waits a random amount of time and then broadcasts a *Request* message including the underload Δ_- and a TTL (Time-To-Live). All intermediate regions that receive the message for the first time decrease the TTL by 1 and forward it. TTL is used to control the number of hops that a message is forwarded. Besides participating in data forwarding, a source region receiving a *Request* message sends back a *Reply* message containing the number of sensors $min(\Delta_-, \Delta_+)$ it allocates and reserves for this request. This is a unicast message transmitted back to the hole that has initiated the request. Once the hole receives *Reply* messages, it computes a movement plan, specifying, for each source region, the number of sensors it has to move. If the number of sensors reserved by the sources is less than or equal to Δ_-, then all of them are included in the movement plan. If the number is greater than Δ_-, then the sensors from the closer sources are added in the movement plan first. This selection criterion helps to minimize the sensors movement distance. The hole then broadcasts a message *MovementPlan* with the same TTL value used in the *Request* message. All intermediate regions that receive the message for the first time decrease TTL by 1 and forward it. The actual sensor movement takes place when a source receives a *MovementPlan* message. After the sensor movement, the source updates Δ_+. There may be cases when not all of the underloaded sensors are filled in the first iteration. In this case, the process is repeated using the expanding ring search mechanism. Thus, in the next iteration the search is performed using $TTL = TTL + \delta$, where δ is a predefined constant. The matching process terminates when all hole regions have filled out their underload values and thus have become neutral regions.

To summarize, the main steps executed by the source regions are as follows:

- Determine the overload value Δ_+.
- If *Request* message received, then reserve $min(\Delta_+, \Delta_-)$ sensors for some specific time, and send back a *Reply* message with the number of sensors $min(\Delta_+, \Delta_-)$ allocated for this request.
- If *MovementPlan* message received that requests n^* sensors from this source, then move n^* sensors to the hole and update $\Delta_+ \leftarrow \Delta_+ - n^*$. If $\Delta_+ = 0$, then change the status to *neutral*.

The main steps executed by the hole regions are as follows:

- Determine the underload value Δ_- and wait a random time.
- Broadcast a *Request* message including Δ_-; use TTL to limit the number of hops.
- If *Reply* messages received, then compute a movement plan including the number of sensors to be moved from each source, giving priority to the closer sources. Broadcast *MovementPlan* message using TTL mechanism to limit the number of hops.
- After the movement phase, update Δ_-.
- If $\Delta_- > 0$, then TTL $=$ TTL $+ \delta$ and go to the step 2. Else if $\Delta_- = 0$, then change status to *neutral*.

13.3.2   Sensor Redeployment Mechanisms for Composite Event Detection

Article [16] considers sensor deployment for composite event detection. The square monitored area is divided into $r \times r$ grids, see Figure 13.14a. To achieve high reliability, the composite event must be k-watched in each grid. Otherwise, a detection breach occurs. The definition for k-*watched atomic/composite event* was introduced in the Section 13.2.2. Given the initial deployment, the goal is to relocate sensors such that to minimize the breach for all regions.

It is assumed that there is at least one sensor in each region. Otherwise, a seed planting mechanism similar to SMART [15] (Section 13.3.1) can be applied. To ensure sensor coverage (a sensor can monitor the whole region), r is chosen such that $r \leq \frac{\sqrt{2}Rs}{2}$, where Rs is the sensing range of the sensor. Each region has a representative which takes care of communication with representatives of the neighbor regions and with the sink. A representative can

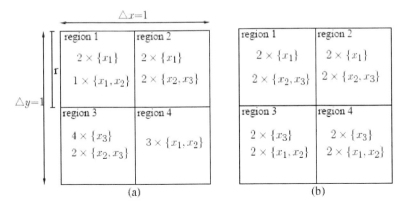

Fig. 13.14 An example of sensor deployment for composite event detection. (a) The initial deployment. (b) Deployment after the sensor relocation.

be chosen according to the sensor's residual energy or contribution. To ensure the communication between adjacent representatives (left, right, top, and bottom), $r \leq \frac{Rc}{\sqrt{5}}$, where Rc is the sensor communication range. A sensor can move to its neighbor regions (left, right, top, and bottom). The paper assumes that sensors have limited mobility capabilities due to the energy constraints. d_{max} denotes the maximum Manhattan distance a sensor can move. M atomic events, $x_1, x_2, \ldots, x_M$ form the composite event. Sensors are equipped with single or multiple sensing components. Figure 13.14 shows an example for $M = 3$. For example, x_1, x_2, and x_3 are temperature, light, and smoke, respectively. For a sensor equipped with temperature and light sensing components, the set $\{x_1, x_2\}$ denotes its sensing ability.

Due to a random initial deployment of the WSN, some regions do not ensure the k-watching property. The *breach* of a region is defined as the maximum gap in achieving the k-watching of an atomic event. $n_i^{x_j}$ denotes the total number of sensing components x_j in region i. The breach of region i is defined as $breach_i = \max(0, k - n_i^{x_1}, k - n_i^{x_2}, \ldots, k - n_i^{x_M})$. In the initial deployment, some regions might have a higher breach, while others might have additional sensors. The problem addressed in [16] is as follows. Given a WSN with N sensors randomly deployed in a square area which is partitioned into R $r \times r$ grids, and given the maximum sensor moving distance d_{max}, design a sensor moving strategy such that to minimize the maximum breach for all the regions in the network, that is $MAX_{i=1,\ldots,R} breach_i = MINIMUM$.

Figure 13.14 shows an example with $k = 2$, $M = 3$, and $d_{max} = 1$. Each of the three atomic events has to be 2-watched in each region and the maximum moving distance of a sensor is 1 (each sensor can only move to a neighboring region). In this example, the monitored area is divided into four regions and there are four sensing combinations $\{x_1\}$, $\{x_1, x_2\}$, $\{x_2, x_3\}$, $\{x_3\}$. Figure 13.14a shows the initial sensor deployment. Region 1 has two sensors $\{x_1\}$ and one sensor $\{x_1, x_2\}$, region 2 has two sensors $\{x_1\}$ and two sensors $\{x_2, x_3\}$, region 3 has four sensors $\{x_3\}$ and two sensors $\{x_2, x_3\}$, and region 4 has three sensors $\{x_1, x_2\}$. The maximum breach in the initial deployment is 2, occurring in regions 1 (no sensing component x_3), 3 (no sensing component x_1), and 4 (no sensing component x_3).

The goal is to relocate sensors such that to minimize the maximum breach for all regions. A possible deployment after sensor repositioning is shown in Figure 13.14b. The moving strategy is: two sensors $\{x_2, x_3\}$ move from region 3 to region 1, one sensor $\{x_1, x_2\}$ moves from region 1 to region 3, two sensors $\{x_3\}$ move from region 3 to region 4, and one sensor $\{x_1, x_2\}$ moves from region 4 to region 3. The total number of sensor movements is 6 and every region has breach 0 (or no breach) after sensors relocations.

One method is a centralized Integer Programming (IP) approach. Let P denote the number of sensing combinations deployed, containing one or more sensing components which can watch atomic events of interest. The objective is to compute the number of sensors with sensing combination $l(1 \leq l \leq P)$ which move from a region s to a region t, $X_{st}^l \in \{0, 1, \ldots, w_s^l\}$, when $d(s, t) \leq d_{max}$. If the Manhattan distance between the regions $d(s, t) > d_{max}$, then $X_{st}^l = 0$. X_{ss}^l represents the number of sensors with sensing combination l which do not leave region s. w_s^l is the original number of sensors equipped with sensing combination l in the region s. The optimal solution is modeled using the following IP formulation:

Minimize $\max_{1 \leq i \leq R} breach_i$

Subject to $\sum_{t=1}^{R} X_{st}^l = w_s^l$ for $1 \leq s \leq R$, $1 \leq l \leq P$

$X_{st}^l = 0$ for $d(s, t) > d_{max}$, $1 \leq s, t \leq R$
     and $1 \leq l \leq P$

Where $X_{st}^l \in \{0, 1, \ldots, w_s^l\}$ for all $1 \leq s, t \leq R$ and $1 \leq l \leq P$

Remarks:

- The objective function asks to minimize the maximum breach for all regions i, $1 \leq i \leq R$.
- The first constraint requires that the number of sensors with sensing combination l that leave or stay in region s to be equal to the total number of sensors with sensing combination l originally in region s, w_s^l.
- The second constraint ensures that no sensor moves from a region s to a region t if the Manhattan distance between them is greater than the maximum distance d_{max}. This constraint guarantees that the total movement distance of a sensor does not exceed d_{max}.

After getting the optimal maximum breach, other IP formulations could be used to optimize the number of movements and the moving distance. The objective function for optimizing the number of movements is to minimize $\sum_{l=1}^{P} \sum_{i=1}^{R} \sum_{j=1}^{R} X_{ij}^l$. For optimizing the moving distance, the objective is to minimize $\sum_{l=1}^{P} \sum_{i=1}^{R} \sum_{j=1, i \neq j}^{R} d(i, j) \cdot X_{ij}^l$. Besides the same two constraints from the previous IP formulation, there is the additional constraint that $breach_i \leq optBreach$, for $1 \leq i \leq R$, where $optBreach$ is the optimal maximum breach.

Another method is using a localized method [16], organized in rounds, that incrementally reduces the maximum breach in the network. The algorithm runs in $k + \sigma$ rounds, where $\sigma > 0$ is a tunable parameter. The regions that initiate the mechanism to reduce their breach in a round are called initiator regions, or simply initiators.

After the initial deployment, the maximum breach in the network is up to k. In the first round, the initiators are the regions with breach k. At the end of this round, depending on the network characteristics, there might be initiators that could not decrease their breach. In the second round, the initiators are the regions with breach k or $k - 1$. More generally, in the hth ($h \leq k$) round, the initiators are the regions with breaches greater than or equal to $k + 1 - h$. In the rounds $k + 1, \ldots, k + \sigma$, the initiators are all the regions with breaches greater than 0. A higher priority is given in reducing higher breaches.

In general, there are two ways to reduce the breach of an initiator. When a region has transferable sensors, then it can directly assign sensors to the initiator. Otherwise, if no sensors can be moved but there are transferable sensing components, the region first tries to get transferable sensors by exchange and then assigns transferable sensors to initiators. The first method has higher priority since fewer sensor movements are involved. In a round, each initiator has zero or more *candidate regions*. A candidate region must have the breach less than the initiator's breach and must satisfy condition 1 and one of the conditions 2 or 3:

1. Being located at a distance less than or equal to d_{max} from the initiator, since sensors' maximum moving distance is upper-bounded by d_{max}.
2. *First class candidate region*: have transferable sensors equipped with one or more of the initiator's *key sensing components* (sensing components with the maximum breach in the initiator). A sensor is *transferable* if moving that sensor from the candidate region will keep the breach of the candidate region less than the initiator's breach. The goal is to minimize the maximum breach in every round, and therefore sensor movements must not generate a new higher breach.
3. *Second class candidate region*: have one or more transferable key sensing components. The candidate may get transferable sensors through exchanging sensors with other regions. In this case, the breach of the candidate will not increase as result of the exchange.

In each round, the algorithm consists of a negotiation process between initiators and their candidates. The process is started by the initiators and after the negotiation process, sensors are moved. In each round, initiators use controlled flooding to send *Request* messages, in an attempt to reduce their breaches. An initiator waits a time T before sending the request in order to reduce collisions, avoiding adjacent neighbors sending requests at the same time. The value of T is computed based on the initiator's breach and a small random number. In general, the higher the breach is, the smaller the value of T is.

The request message generated by an initiator i has the format: *Request* $(RID = i, n_i^{x_1}, n_i^{x_2}, \ldots, n_i^{x_M}, breach, TTL = d_{max})$. *RID* is the initiator's identifier. $\{n_i^{x_1}, n_i^{x_2}, \ldots, n_i^{x_M}\}$ are the numbers of each sensing component the initiator has and they also indicate which types of sensing components are needed. *breach* is the breach of the initiator and TTL (Time-To-Live) is used to control the number of hops the message is forwarded.

When a second class candidate receives the request, it asks for exchanging sensors. Consider the case when a request asks for the key sensing component $\{x_1\}$ and the candidate has a sensor $\{x_1, x_2\}$ which can not be directly assigned to the initiator since sensing component x_2 may still be needed for the candidate. If the candidate region can exchange the sensor $\{x_1, x_2\}$ for one sensor $\{x_1\}$ and another sensor $\{x_2\}$, then it can give the initiator the transferable sensor $\{x_1\}$. The candidate region asks for an exchange by attaching the fields: *{RID, keyComp, otherCompList, EXCHANGE}* to the end of the request message, where *RID* is the identifier of the exchange requestor, *keyComp* is the key sensing component, and *otherCompList* is the list of the other sensing components in the sensing combination and *EXCHANGE* shows the type of the movement. The candidate region decreases TTL in the request by 1 and forwards the message if TTL ≥ 1.

When a first class candidate receives a request message, it follows the following steps to decide whether it will give one or more sensors to the initiators.

1. A region may receive more than one request messages. The candidate region i computes a priority for each request it receives and sorts them in decreasing order, assigning first sensors to the initiators with higher priorities. For a request from the initiator region j, the priority is the initiator's breach value, taken from the request message. The Manhattan distance between these two regions, which can be computed from the TTL value in the request message, is used to be the second criterion to break the tie. A shorter distance has higher priority.

2. The contribution $\varphi_i = \frac{hf_i}{sc_i}$ of each transferable sensor i is computed, where hf_i is the number of helpful sensing components in the transferable sensor i and sc_i is the total number of sensing components sensor i has. Note that the transferable sensor must contain at least one key sensing component of the initiator. Otherwise, it can not reduce the breach of the initiator. For example, if the initiator's request message has breaches for the sensing components x_1 and x_2, then the contribution of a sensor $\{x_1, x_3\}$ is $\frac{1}{2}$, since only $\{x_1\}$ is helpful for this request and the sensor has 2 sensing components in total. The contribution of a sensor $\{x_1\}$ is 1.

3. A candidate region addresses the requests in the sorted order. It computes the contribution of each transferable sensor, sorts them in decreasing order, and assigns the first $\lfloor \delta \cdot N_{transfer} \rfloor$ sensors to the request. δ is an input parameter and $N_{transfer}$ is the number of transferable sensors.

4. After the decision is made, the candidate sends back *Reply* messages to the initiators to whom sensors have been assigned. A *reply* message has the form *Reply* (RID_1, RID_2, *sensorList*, *breach'*, *dist*, *GIVE*), where RID_1 is the candidate region identifier and RID_2 is the initiator's identifier. *sensorList* is the list of sensors which can be assigned to the initiator. The breach of the candidate may change due to the assignment. *Breach'* is the new breach and *dist* is the distance between the candidate region and the initiator region. *GIVE* means the sensors come from a first class candidate. The reply is a unicast message sent along the reverse path established when the request message was sent.

5. If the TTL in the received request message is greater than 1, it decrements the TTL by 1 and forwards it. If the received request message contains an exchange request whose requested sensing components have been assigned by the candidate region, then the exchange information is removed from the request message.

When an intermediate region which is not a candidate (non-candidate region) receives a request message, it checks the attached exchange information if there is any. If it can make the exchange, it sends back a *Reply* message to the exchange requestor. The message has the form *Reply* (RID_1, RID_2, *keyComp*, *otherCompList*, *EXCHANGE*), where RID_1 is the identifier of the region sensing the reply message, RID_2 is the identifier of the exchange requestor, *keyComp* and *otherCompList* are the sensing components that can be exchanged. Then the request message is updated (TTL is decremented by 1 and the exchange information is removed) and forwarded if TTL ≥ 1.

When the exchange requestor receives the reply message, it updates the reply message with its own transferable sensor information and forwards it to the corresponding initiator.

An initiator may receive multiple reply messages from several candidate regions. It follows the following steps to decide which sensors to take from the first class candidates and then, if the breach is still greater than 0, the initiator considers taking sensors from second class candidates.

1. It computes the contribution $\varphi_i = \frac{hf_i}{sc_i}$ for each sensor i in the *sensorList* of the message.
2. The initiator considers sensors from *sensorList* of the candidates in increasing order of the *breach'* values first and then in the decreasing order of the sensor's contribution to break the tie.
3. The initiator considers taking sensors as long as its breach can be decreased and is greater than 0.
4. The initiator sends one *ACK* message indicating the movement plan (number and types of sensors it will take from each candidate). The *ACK* message is sent using localized flooding with TTL equal to the distance between the initiator and the farthest candidate with sensors in the movement plan. The sensor reservation made by other candidates will expire if they are not included in an *ACK* message.

When first class regions receive *ACK* messages, sensors are actually moved to the initiator. When second class regions receive *ACK* messages, they first exchange sensors and then move the corresponding transferable sensors to the initiator.

13.4 Summary of Chapter

In this chapter, we introduce the definition of atomic event detection and composite event detection. Sensors equipped with different numbers and types of sensing components can detect an atomic event independently or they can cooperate to detect a composite event. We focus on the sensor scheduling and sensor redeployment mechanisms in wireless sensor networks. Sensor scheduling mechanisms put redundant sensors to sleep to save energy and prolong network lifetime. A set of active sensors is selected to perform sensing and data relaying while the other sensors go to sleep. Sensors work alternatively to prolong network lifetime. Sensor redeployment mechanisms use sensor mobility to relocate sensors to improve the initial deployment. The goal is to achieve load balance, a reliable surveillance, or to prolong network lifetime. For the composite event detection, sensor scheduling and redeployment problems become more complicated. Several centralized

Integer Programming approaches, distributed approaches, and localized algorithms have been studied in this chapter.

13.5 Acknowledgements

This work was supported in part by the NSF grant CCF 0545488.

Problems

Problem 13.1

What is a composite event? Show an example of composite event detection.

Problem 13.2

Extend the Geographical Adaptive Fidelity (GAF) algorithm [23] to mobile networks, where nodes can move and active nodes may leave their original grids.

Problem 13.3

Apply the algorithm in [12] to find out the detection set for the scenario in Figure 13.15. The composite event is the combination of $\{x_1, x_2, x_3\}$ and every atomic event must be 3-watched ($k = 3$).

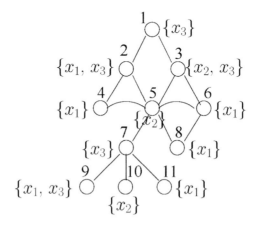

Fig. 13.15 The initial graph for the problem 13.3.

Problem 13.4

Extend the flip-based sensor deployment method [4] to the following cases:

1. $R = d$, $F > d$, $C = 1$.
2. $R = d$, $F > d$, $C = n$.
3. $R > d$.

where R is the region size, F is the maximum distance a sensor can flip, C denotes the flip choices, and d is the basic flip unit. Describe how to construct the virtual graphs for the above cases.

Problem 13.5

Design an effective seeds planting mechanism for the SMART algorithm [15]. Explain how the clusters with more than one sensor can effectively send sensors to the clusters without any sensors, such that there is at least one sensor in each cluster before the scan method is applied.

Problem 13.6

Articles [17] and [22] discuss the non-uniform sensor density to prolong network lifetime. Compare the uniform and non-uniform sensor distributions and compute the improvement in network lifetime obtained by deploying a non-uniform sensor distribution.

References

[1] I. Akyildiz, W. Su, Y. Sankarasubramaniam, and E. Cayirci, "Wireless sensor networks: a survey," Computer Networks, 38(4), pp. 393-422, 2002.

[2] Q. Cao, T. F. Abdelzaher, T. He, and J. A. Stankovic, "Towards optimal sleep scheduling in sensor networks for rare-event detection", IPSN 2005: 20–27.

[3] M. Cardei, M. Thai, Y. Li, and W. Wu, "Energy-efficient target coverage in wireless sensor networks", IEEE INFOCOM 2005, March 2005, Miami, USA.

[4] S. Chellappan, X. Bai, B. Ma, D. Xuan, and C. Xu, "Mobility limited flip-based sensor networks deployment", IEEE Transactions of Parallel and Distributed Systems, 18(2), pp. 199–211, 2007.

[5] K. Dantu, M. H. Rahimi, H. Shah, S. Babel, A. Dhariwal, and G. S. Sukhatme, "Robomote: enabling mobility in sensor networks", IPSN 2005, pp. 404–409, 2005.

[6] S. Gao, C. T. Vu, and Y. Li, "Sensor scheduling for k-coverage in wireless sensor networks", MSN 2006: 268–280.

[7] J. Jeong, S. Sharafkandi, and D. Du, "Energy-aware scheduling with quality of surveillance guarantee in wireless sensor networks," the 2nd ACM/SIGMOBILE Workshop on

Dependability Issues in Wireless Ad Hoc Networks and Sensor Networks (DIWANS), September 2006.

[8] A. V. U. P. Kumar, A. M. Reddy V., and D. Janakiram, "Distributed collaboration for event detection in wireless sensor networks", MPAC, 2005.

[9] U. Lee, E. O. Magistretti, B. O. Zhou, M. Gerla, P. Bellavista, and A. Corradi, "Efficient data harvesting in mobile sensor platforms", PerCom Workshops, pp. 352–356, 2006.

[10] C. Liu, K. Wu, Y. Xiao, and B. Sun, "Random coverage with guaranteed connectivity: joint scheduling for wireless sensor networks", IEEE Trans. Parallel Distrib. Syst. 17(6): 562–575 (2006).

[11] S. Olariu and I. Stojmenovic, "Design guidelines for maximizing lifetime and avoiding energy holes in sensor networks with uniform distribution and uniform reporting", IEEE INFOCOM, 2006.

[12] C. V. Vu, R. A. Beyah, and Y. Li, "Composite event detection in wireless sensor networks", 26th IEEE IPCCC, 2007.

[13] B. Wang, K. C. Chua, V. Srinivasan, and W. Wang, "Scheduling sensor activity for point information coverage in wireless sensor networks", WiOpt 2006: 353–360.

[14] G. Wang, G. Cao, and T. F. LaPorta, "Movement-assisted sensor deployment", IEEE Transactions on Mobile Computing, 5(6), pp. 640–652, 2006.

[15] J. Wu and S. Yang, "SMART: a scan-based movement-assisted sensor deployment method in wireless sensor networks", INFOCOM'05, pp. 2313–2324, 2005.

[16] Y. Yang and M. Cardei, "Sensor deployment for composite event detection in mobile WSNs", International conference on Wireless Algorithms, Systems and Applications (WASA), October 2008.

[17] Y. Yang and M. Cardei, "Movement-assisted sensor redeployment scheme for network lifetime increase", MSWIM'07, October 2007.

[18] Y. Yang, A. Ambrose, and M. Cardei, "Coverage for composite event detection in wireless sensor networks", under review.

[19] T. Cormen, C. Leiserson, R. Rivest, and C. Stein, Introduction to algorithms, in MIT Press, 2001.

[20] A. V. Goldberg, "An efficient implementation of a scaling minimum-cost flow algorithm", in J. Algorithms 22, 1997.

[21] M. Marta, Y. Yang, and M. Cardei, "Energy-efficient composite event detection in wireless sensor networks", International Conference on Wireless Algorithms, Systems and Applications (WASA), August 2009.

[22] M. Cardei, Y. Yang, and J. Wu, "Non-uniform sensor deployment in mobile wireless sensor networks", The IEEE Intl. symposium on a World of Wireless, Mobile and Multimedia Networks (WoWMoM'08), June 2008.

[23] Y. Xu, J. Heidemann, and D. Estrin, "Geography-informed energy conservation for ad hoc routing," in the ACM/IEEE Intl. Conf. on Mobile Computing and Networking (Mobi-Com'01), July 2001.

[24] J. G. McNeff, "The global positioning system", IEEE Transactions on Microwave Theory and Techniques, Mar. 2002; Vol. 50, pp. 645–652.

[25] T. He, C. Huang, B. M. Blum, J. A. Stankovic, and T. Abdelzaher, "Range-free localization schemes for large scale sensor networks", ACM Mobicom, 2003.

[26] L. Hu and D. Evans, "Localization for mobile sensor networks", ACM Mobicom, 2004.

[27] S. Singh, C. S. Raghavendra, and J. Stepanek, "Power efficient broadcasting in mobile ad hoc networks", Proceedings PIMRC'99, 1999.

14

Classifier Combination

Melanie Osl*, Christian Baumgartner* and Stephan Dreiseitl[†]

*University for Health Sciences, Medical Informatics and Technology, Austria
[†]Upper Austria University of Applied Sciences, Austria

In this chapter we give a survey of the combination of classifiers. Section 14.1 to 14.3 briefly describe basic principles of machine learning and the problem of classifier construction. Section 14.4 reviews several approaches to generate different classifiers as well as established methods to combine different classifiers. Then, from Section 14.5, we introduce our novel approach to assess the appropriateness of different classifiers based on their characteristics for each test point individually.

14.1 Machine Learning

The field of machine learning seeks to develop methods to learn from data. Therefore, computational and statistical methods are applied, which connects machine learning closely to statistics and theoretical computer science.

Machine learning is divided into supervised and unsupervised learning. In supervised learning, a set of data points with known prediction values is given. The objective is to model the conditional distribution of the prediction values given the data points in order to make predictions about future data. According to the characteristics of the predicted values, supervised learning can be further classified into regression and classification. In regression, the prediction values are continuous. An example for a regression task is to predict a patient's blood pressure based on clinical data and lab results. In classification, the

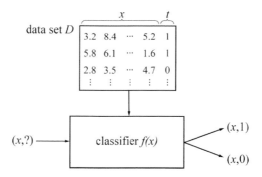

Fig. 14.1 Illustration of a classifier system (*t* is a class label, *x* is a vector of a attribute valuse.)

predicted values are distinct classes. Examples for classification tasks are spam filtering, medical decision support, or fraud detection [1]. If no prediction value is given in the data, unsupervised learning methods are applied with the objective of grouping data points according to similar characteristics, e.g. to identify groups of interacting genes. In the following we concentrate on classification, especially on two-class classification problems, as illustrated in Figure 14.1.

14.2   Two-Class Classification Problem

Formally, the given data set D consists of data points (x, t), where x is an n-dimensional vector of attributes and $t \in \{0, 1\}$ is a class label. From this data set a model $f(x)$ of the conditional probability $P(t|x)$ is learned. The function $f(x)$ is also called classifier. In order to classify an unseen data point x_k the classifier computes $f(x_k)$ and assigns the data point to class $t_k = 0$ if $f(x_k) \geq 0.5$ and $t_k = 1$ if $f(x_k) < 0.5$. Some of the most popular classifiers are logistic regression, classification trees, artificial neuronal networks, k-nearest neighbors and support vector machine.

Logistic Regression

Logistic regression searches for an optimal linear separation between two classes. It models the posterior probabilities of the classes in terms of a logit transformation to ensure that the probabilities sum up to one and remain in the range of 0 to 1. The parameters of a logistic regression are optimally adjusted by maximum likelihood [2].

Classification Tree

Classification trees classify data points by sorting them down a tree where each node corresponds to a test on some attribute of the data point, and each branch corresponds to one of the possible values of this attribute. To build up the tree, all attributes are evaluated using a statistical test. The one which classifies the data set best is chosen to be placed at the root node. A descendant node for each possible value of the attribute is created, and the data points are sorted in. The process is repeated for each descendant node using the data points associated with this node [3].

Artificial Neural Network

An artificial neural network classifier consists of a set of artificial neurons organized in layers. The neurons of one layer serve as input for the neurons of the next layer and are therefore connected to each other. Each neuron takes a vector of weighted inputs, calculates a linear combination of these inputs and outputs a logistic transformation of this linear combination. The weights are adapted to minimize the deviation of the final output of the network to the expected output of the network [4].

Support Vector Machine

A support vector machine extents the solution of an optimal linear separation problem to the situation where classes are not linearly separable. It produces nonlinear decision boundaries by constructing a linear boundary in a transformed attribute space. An optimal separation is found by maximizing the so-called margin, the largest possible distance between the separation hyperplane and the data points on either side. The points that lie on the margin are called support vectors and represent, as a linear combination, the solution of a convex optimization problem [5].

k-Nearest Neighbors

A k nearest neighbors classifier searches the training set for the k closest neighbors to a given test point according to a certain distance measure (e.g. Euclidean distance). Commonly, the majority class of the neighbors is assigned to the test point, but also a probability estimate given by the frequency of the test points for which $t_k = 1$ can be returned [6].

14.3 Classifier Evaluation

To assess the quality of a classifier, a common approach is to split the data set into a *training set* and a *test set*. The training set is used to build the classifier. Then the classifier is evaluated by measuring the percentage of correctly classified test points, referred to as *accuracy* of the classifier. However, this approach is not feasible for small data sets. To overcome this problem, the process of training and testing is repeated on different splits of the data. In *k-fold cross-validation* the data set is split into k parts, where k times a different part is used for testing and the remaining $k-1$ parts are used for training. The k accuracies obtained for the k folds are finally averaged. If the number of folds is chosen so that the test set of each fold consists of only one test point, the validation procedure is called *leave-one-out cross-validation*.

However, for data sets with highly skewed class distributions, accuracy was shown to be an inappropriate quality measure [7]. For example, a two class data set with 70 data points in one class and 30 data points in the other class would be classified with an accuracy of 70 percent by an uninformed classifier that predicts only based on class prevalence. An alternative measure to accuracy for assessing the quality of a classifier is the *area under the ROC curve (AUC)*. The AUC for the example above is 0.5, which corresponds to the true situation that the classifier assigns classes to test points only randomly. Thus, AUC has found widespread use as the measure of choice for evaluating classifier performance in the machine learning literature [8].

Given m classifier outputs a_i for data points of class 0, and n classifier outputs b_j for data points of class 1, the AUC estimate $\hat{\theta}$ of the classifier can be shown to be equivalent to

$$\hat{\theta} = \frac{1}{m \cdot n} \sum_{i=1}^{n} \sum_{j=1}^{m} 1_{a_i < b_j} \qquad (14.1)$$

where $1_{(\cdot)}$ denotes the boolean indicator function that returns one if its argument is true, and zero otherwise [9]. It is immediately obvious that $\hat{\theta}$ is an unbiased estimator of the parameter $\theta = P(A < B)$, the probability that a randomly chosen element A of class 0 is ranked lower than a randomly chosen element B of class 1. The value of $\hat{\theta}$ can thus be used as an assessment of a classifier's discriminatory power (how well it can separate two classes). A perfect classifier has $\hat{\theta} = 1$, indicating that there exists a threshold along which both classes can be separated without error. An uninformed classifier that performs no better than chance has $\hat{\theta} = 0.5$.

14.4 Ensemble Classification

For the data set at hand, traditionally the best classifier is selected on the basis of a quality evaluation. Another possibility is to combine the collected classifiers to construct a classifier that performs better than any of the base classifiers [10, 11]. Therefore the base classifiers have to be diverse but also comparable.

A consistent ensemble of classifiers can be generated e.g. by different initializations, different parameter choices or different architectures of the same classifier. More successful due to the fact that the classifiers are independent of each other is to generate different classifiers from different training sets. Well known examples are *bagging* [12] and *boosting* [13, 14]. Bootstrap aggregating (short "bagging") creates n times a set of m data points by sampling uniformly with replacement from the original data set. Thus some data points may be included in the sample multiple times whereas other data points do not appear at all. In boosting, the single training sets differ systematically as the weights for each data points are changed based on the previous classification. Another way to force diversity is to build a set of classifiers on different sets of features [15, 16], e.g. speech and image features in identification problems. Manipulating the output of the training data is particularly useful if the number of output classes is large. Then new learning problems can be constructed by randomly dividing the output classes into a smaller number of subsets of classes [17]. Of course a combination of different classifiers trained on the same features and by the same training set is also possible.

Traditionally the results of the different base classifiers, which can be class labels or class posterior probabilities, are combined by combinatorial functions. The majority voting is the simplest of all combinatorial functions. The class which is most often predicted by the base classifiers is selected. Using functions like sum, product, average or median to combine the results of the base classifiers presumes that the base classifiers are independent of each other, which is hardly ever the case. However, if an ensemble consists of similar base classifiers with independent noise behavior, the errors are averaged out by the summation. Statistical combination methods, e.g. Bayesian combination

[18], are another possibility to build an ensemble classifier. Finally, learning a meta-classifier based on the outputs of different classifiers, called stacking [19], can also be seen as a decision logic for a multi-classifier system.

Beside training diverse base classifiers in parallel and combining them afterwards, a multi-classifier system can also be built of cascading base classifiers or hierarchically. Cascading classifiers pass the classification results from classifiers to classifiers (which take them as input) until the final output is obtained through the final classifier in the chain. However, the major disadvantage of this approach is that the later classifiers are unable to correct mistakes made by the earlier classifiers. The most prominent hierarchical classifier systems are decision trees. A particular advantage of hierarchical classifier systems is that they allow to introduce error checking.

Like experts in a committee, also classifiers in an ensemble have different areas of expertise. By assigning equal weights to each of them at combination these differences in skills are neglected. Therefore, weighted combinations methods are used, where the weights reflect the importance or significance assigned to the classifier. A more objective measure reflecting the confidence into a classifier is its quality found by an independent evaluation set.

14.5   Classifier Characteristics

In contrast to the methods mentioned above, our approach assesses the confidence into different classifiers based on their characteristics and for each test point individually. Thus the combination of the classifiers emphasizes their strengths while diminishing their weaknesses. We elaborate this concept to combine logistic regression and k-nearest-neighbor classification.

Logistic regression (LR) models have a long history as the primary tool for supervised classification problems [20, 21, 22]. For an n-dimensional data set containing two classes, the logistic regression model depends on an n-dimensional parameter vector β and a scalar β_0 in the functional form

$$P(t = 1 | x, \beta, \beta_0) = \frac{1}{1 + e^{-(\beta \cdot x + \beta_0)}} \qquad (14.2)$$

Here, $x \in \mathbb{R}^n$ is alternatively called *covariate vector* or *input vector*, and $t \in \{0, 1\}$ is the class label. The LR model thus provides the probability that a given vector x belongs to class 1. LR models are examples of discriminative models, because they offer a functional representation of a discriminatory line — e.g., where $P(t = 1|x, \beta, \beta_0) = 0.5$ — that separates the two classes in a data set. The LR model output for a data point x depends on the distance of x from the discriminatory line: data points x that are far from the discriminatory line have model outputs close to zero (for x on one side of the line) and close to one (for x on the other side). Data points close to the discriminatory line have outputs close to 0.5. Generally, the parameters of an LR model are estimated by maximum likelihood, i.e., by minimizing the negative log likelihood

$$\sum_{k=1}^{m} \log(1 + e^{(-2t_k+1)(\beta \cdot x_k + \beta_0)}) \tag{14.3}$$

for an m-element data set of input/label pairs (x_k, t_k). LR models are linear in the parameters, and can thus only model separating $(n-1)$-dimensional hyperplanes in n-space.

With the same notation as above, a kNN classifier output can be calculated as

$$P(t = 1|\{x_i, t_i\}) = \frac{1}{k} \sum_{j=i_1}^{i_k} t_j \tag{14.4}$$

where $i_1, \ldots, i_k$ denote the indices of the k points in $x_1, \ldots, x_m$ that are closest (usually in Euclidean distance) to the given data point x. The probability estimate for x is thus given by the local posterior probability in the vicinity of x.

LR models occupy one end of the spectrum spanned by the bias variance tradeoff [23]: Due to their linear nature, LR models have high bias, but little variance, compared to other (nonlinear) machine learning algorithms. On the other end of the spectrum, we can find so-called memory-based classifiers such as k-nearest neighbors (kNN). These classifiers are not models in the strict sense of the word, because they do not build a functional or probabilistic representation of the data. For kNN, the data is the model, implying that there is high variability (and low bias) between different (finite) data samples drawn from the same data distribution. Thus, these two classification methodology complement each other well and combine the advantages of both rigid model structure (by the LR component) and local flexibility (by the kNN component) in one classification structure. The goal of this combination is to obtain better classification performance.

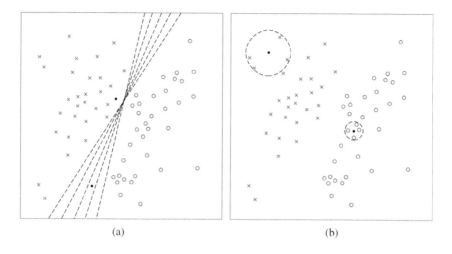

(a)   (b)

Fig. 14.2 Illustration of the topological dependency of LR (left) and kNN (right) appropriateness.

It is important to note that the appropriateness of LR and kNN depends on the local topology around the data point to be classified. An example for two different degrees of appropriateness for LR and kNN is shown in Figure 14.2(a) and 14.2(b), respectively. In Figure 14.2(a) it can be seen that there are several possibilities to separate the two classes, depicted by x and o. The different discriminatory lines are illustrated as dashed lines. A data point at the position of the solid dot near the center of the figure has a similar prediction value for each discriminatory line. In contrast, the prediction for the solid dot in the lower third of the figure differs for the different discriminatory lines. Hence, the application of LR is more appropriate for the upper data point than for the lower one. How appropriate the application of kNN for a data point is depends on the distance to its k nearest neighbors. In Figure 14.2(b) the k nearest neighbors (in this example is $k = 4$) of the data point at the position of the solid dot near the center of the figure are closer than the k nearest neighbors of the data point at the position of the solid dot in the upper left corner of the figure. Hence, the application of kNN is more appropriate for the lower data point than for the upper data point.

Our approach to combining LR and kNN models is thus by individually weighting the contributions of both models for each data point. We assign a larger factor to the model we consider more appropriate for a data point, and a smaller factor to the other model. The methodology to assess the appropriateness of a model for a given data points is different for LR and kNN.

14.6 Classifier Combination

For LR, given a training set, we generate a subset of this training set containing 90% of the data points, and train an LR model on this reduced set. We do this 10 times for a total of 10 LR models that are all slightly different. We then generate the output of each of these LR models on every element of the training set, thus obtaining a standard deviation of model outputs for every data point. Although these standard deviations are likely smaller than those obtained from an independent test set, their average can nevertheless serve as a benchmark that allows us to judge how appropriate the LR models are. The reasoning is that a data point with large standard deviation over all LR models is one for which there is large variability between the models (high variance), and the model output should not be relied upon as much as for a data point with lower variance. The mean over all standard deviations on the training set is used to provide a scale information, as otherwise there is no way to know what constitutes a large or small standard deviation, respectively.

The appropriateness of kNN probability estimates is based on the distances of the k nearest neighbors to a data point. If these distances are large, we consider the probability estimate to be less reliable than if the distances are small. This is because for larger distances, the neighbors are not as close, and the kNN output is not as good an estimate of the local probability as in the case when the neighbors are in the immediate vicinity of the data point. To obtain an average distance value that can serve as a benchmark in the same way as the mean standard deviation for LR models, we calculate the distance of all data points in the training set to their k nearest neighbors. We then take the average of this distribution of distances, and judge the appropriateness of the kNN contribution by relating the kNN distance of a particular data point to the mean distance.

The precise manner in which LR and kNN estimates are combined is as follows: Let $\bar{\sigma}$ denote the mean standard deviation of LR model outputs and $\bar{d}$ the mean distances in the kNN part, both as described above. Also, for the ith data point in the test set, let lr_i denote the LR model output (the mean of all 10 trained models), σ_i the standard deviation of the 10 values, nn_i the

kNN probability estimate, and d_i the distance to the k nearest neighbors. We measure the (in)appropriateness of the two components LR and kNN for this data point as

$$app_{LR} = \frac{\sigma_i}{\bar{\sigma}} \quad \text{and} \quad app_{kNN} = \frac{d_i}{\bar{d}} \qquad (14.5)$$

respectively. Note that a *high* value of one of these parameters means that the respective model is not appropriate, as the point displays above average standard deviation or distance. The contribution of each model is then weighted with the relative inappropriateness of the *other* model: This means that a point for which the LR output is highly inappropriate will assign most of its weight to the kNN component, and vice versa.

The heuristic to calculate a posterior class membership probability by combining all these pieces of information is

$$P(t = 1|x_i, lr_i, nn_i, app_{LR}, app_{kNN})$$
$$= \frac{app_{kNN}}{app_{LR} + app_{kNN}} lr_i + \frac{app_{LR}}{app_{LR} + app_{kNN}} nn_i \qquad (14.6)$$

Note that by using a convex combination of the two model contributions, we again obtain a probability estimate that is in the range of 0 to 1.

14.7 Experimental Setup

We validated our approach on two different data sets: one synthetically generated data set that allows us to graphically demonstrate the feasibility of our method, and one real-world data set containing clinical information about patients with pigmented skin lesions.

Synthetic data

The synthetic data set, of which the training set is depicted in Figure 14.3, consists of two samples drawn from two different Gaussian mixture models with different class means and covariance matrices. Class 0 comprises 400 data points shown as o; class 1 consists of 600 data points marked by x. The dashed line shows the class separation by the logistic regression model at a threshold of $p = 0.5$. The two highlighted o points are used to provide, in Section 14.8, two examples of how our algorithm weights the contributions of LR and kNN differently, based upon our assessment of the appropriateness of the two components.

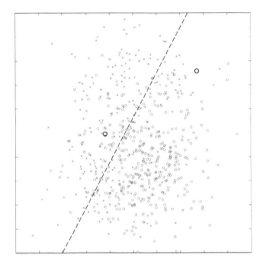

Fig. 14.3 Illustration of the synthetic data set.

Melanoma data

The real-world data set was collected at the pigmented lesion unit of the Department of Dermatology at the Medical University of Vienna, Austria. It is divided into the three classes *common naevi*, *dysplastic naevi*, and *melanomas*. The class distribution (gold standard) is 1290 patients for which a suspicious lesion was diagnosed as a common naevus, 224 patients with dysplastic naevi, and 105 patients with melanomas. For each of the 1619 patients, five clinical data items were recorded: the skin type according to Fitzpatrick, personal and family history with regard to melanoma, whole body naevus count, and skin damage due to sun exposure. In our experiments, we investigated the dichotomous problem of distinguishing patients with common naevi and dysplastic naevi from those with melanomas, based only on their clinical information. We had previously investigated the performance of a number of machine learning algorithms on an extension of this data set, which also included lesion features obtained by dermoscopy (epiluminescence microscopy) [24].

The data sets were split into training and test set. We used 60% of the data to train the algorithm, and 40% to test it. For the synthetic data set, the training set contained a total number of 240 data points from class 0, and 360 data points from class 1. The test set contained the remainder: 160 data points from class 0, and 240 data points from class 1. For the melanoma data set, the training set consisted of 909 data points from class 0 (common naevi and dysplastic naevi) and 63 data points from class 1 (melanomas). The test set consisted of the remaining 605 data points from class 0 and 42 data points from class 1.

To consider the influence of different data set splits on our algorithm, we applied our approach on 20 different splits. This means that we randomly generated 20 different training and test sets with class distributions as described above. We calculated the area under the ROC curve for all of these splits and for all three algorithms (LR and kNN considered separately, and combined by our method). Our final performance numbers are the means of AUCs over all data set splits.

14.8   Results and Discussion

In Table 14.1, we compare our results with those of LR and kNN applied separately. We fixed the value of $k = 10$ in our k-nearest-neighbor calculations. The AUC values of LR for each of the 20 splits were calculated by averaging the posterior class membership probabilities of all 10 models generated for assessing the variability of LR model outputs. The entries in Table 14.1 are the average (over 20 splits) of these averages (over 10 models). Besides these mean AUC values, we also report the standard deviations (SD) for LR, kNN, and our combination method of combination (denoted by "comb" in the table) on the synthetic and biomedical data set. It can be seen that our method achieves a higher classification performance for both data sets.

The advantage and novelty of our approach to data classification can be attributed to the heuristic method we use to combine LR model outputs and kNN local probability estimates. To illustrate this point, we consider the two data points highlighted in Figure 14.3. Both of the points belong to class 0. In this figure, the LR model classifies all points to the left of the dashed discriminatory line as belonging to class 1, and all points to the right of the

Table 14.1 Performance comparison.

	synthetic data		melanoma data	
	AUC	SD	AUC	SD
LR	0.7748	0.0134	0.6777	0.0327
kNN	0.7763	0.0116	0.7479	0.0355
comb	0.7940	0.0116	0.7850	0.0386

line as belonging to class 0. In the following, we will use the notation of Equations 14.1 and 14.2. For the training set shown in Figure 14.1, the average standard deviation over the 10 LR models was $\bar{\sigma} = 0.0071$, and the average distance of a data point to its $k = 10$ nearest neighbors was $\bar{d} = 3.8872$.

The first point (on the left side of the LR discriminatory line) was misclassified by the LR model, since it is on the wrong side of the line (the LR model output was $lr_i = 0.6282$). In contrast, the kNN probability estimate was $nn_i = 0.2$, which is closer to the true value of 0. The LR variability estimate of this data point was $\sigma_i = 0.0067$, and the kNN distance was $d_i = 3.1074$. The two measures of appropriateness, as defined in Equation (14.5), were thus $app_{LR} = 0.9437$ and $app_{kNN} = 0.7994$. This means that the LR model is less appropriate than the kNN part; the LR component therefore only contributes 0.4586 to the combined model, as calculated by the weighting factors in Equation (14.6). The remaining weight of 0.5414 is assigned to the kNN contribution, resulting in an overall combined model output of 0.3964. By weighting the correct kNN contribution more heavily than the incorrect LR contribution, our method was able to correct the error of the LR model.

The second data point (on the right side of the discriminatory line) was classified correctly by the LR model, with a value of $lr_i = 0.2577$. For this point, the kNN estimate is incorrect, with $nn_i = 0.8$. Because of the values $\sigma_i = 0.0114$ and $d_i = 8.1644$, which result in appropriateness estimates of $app_{LR} = 1.6056$ and $app_{kNN} = 2.1003$, the incorrect contribution of the kNN component is weighted less heavily than the correct contribution of the LR component. The combined model estimate was 0.4927. The kNN contribution was downweighted, and not large enough to cause an incorrect classification at the $p = 0.5$ LR threshold.

By weighting the contributions of both classifiers for each data point we consider their characteristics. Being limited to linear discriminatory hyperplanes, LR models have less variance than kNN classifiers. However, with increasing distances of the k nearest neighbors, the class prediction of a kNN classifier becomes less accurate. In this case, the LR output is considered to be more appropriate. On the other hand, if there is (relatively) high variability between the 10 LR models we generate, we assess the local probability estimate of kNN more appropriate.

Our experiments showed that our method of combining LR and kNN classifiers achieves a higher classification performance than the individual classifiers, both on a synthetically generated data set and on a biomedical data set. The improvement on the biomedical data set was higher than the improvement on the synthetic data set. We speculate that this may be due to the data set characteristics: The synthetic data set consists only of real-valued features, whereas the melanoma data set contains real, ordinal and categorical features. We presently investigate which data set characteristics may have an influence on the success of our method.

Possible directions for future research lie in the investigation of different heuristics for assessing the variability of LR models (possibly by bootstrapping methods), and in alternative methods to improve the kNN component of the model, either by applying adaptive kNN methods [25], or by weighting the distance calculations by the LR β coefficients.

14.9   Summary

In this chapter we presented the context for combining classifiers. We reviewed several approaches to generate different classifiers as well as established methods to combine different classifiers. Then we introduced a novel ensemble method to weight different classifiers based on their characteristics and for each test point individually. We elaborate this approach by combining a rigid LR model with the local flexibility of a kNN classifier. The individual weighting of both models for each given data point exploits the strength of LR, having little variance, as well as the strength of kNN, having low bias. Our experimental results on a synthetic and a biomedical data set confirm the feasibility and power of our approach. The combination of the models achieves, on both data sets, a higher classification performance than the individual models.

Acknowledgements

This work was supported by the Austrian Genome Program (GEN-AU), project Bioinformatics Integration Network (BIN III).

References

[1] Christian Baumgartner, Christian Böhm, Daniela Baumgartner, *et al.*, "Supervised machine learning techniques for the classification of metabolic disorders in newborns," Bioinformatics, vol. 20, pp. 2985–2996, 2004.

[2] S. Dreiseitl and L. Ohno-Machado, "Logistic regression and artificial neural network classification models: A methodology review," Journal of Biomedical Informatics, vol. 35, pp. 352–359, 2002.

[3] J. Quinlan, C4.5: Programs for Machine Learning, Morgan Kaufmann Publishers, 1993.

[4] C. Bishop, Neuronal Networks for Pattern Recognition, Oxford University Press, 1995.

[5] N. Cristianini and J. Shawe-Taylor, An Introduction to Support Vector Machines, Cambridge University Press, 2000.

[6] T. Mitchell, Machine Learning, McGraw Hill, 1997.

[7] F. Provost, T. Fawcett, and R. Kohavi, "The case against accuracy estimation for comparing induction algorithms," Proceedings of the 15th International Conference on Machine Learning, 1998, pp. 445–453.

[8] A. Bradley, "The use of the area under the ROC curve in the evaluation of machine learning algorithms," Pattern Recognition, vol. 30, pp. 1145–1159, 1997.

[9] D. Bamber, "The area above the ordinal dominance graph and the area below the receiver operating characteristic graph," Journal of Mathematical Psychology, vol. 12, pp. 387–415, 1975.

[10] K. Chen, L. Wang, H. Chi, "Methods of combining multiple classifiers with different features and their application to text-independent speaker identification," International Journal of Pattern Recognition and Artificial Intelligence, vol. 11, pp. 417–445, 1997.

[11] Y.S. Huang, C.Y. Suen, "A method of combining multiple experts for the recognition of unconstrained handwritten numerals," IEEE Transactions on Pattern Analysis and Machine Intelligence, vol. 17, pp. 90–94, 1995

[12] L. Breiman, "Bagging predictors," Machine Learning, vol. 24, pp. 123–140, 1996.

[13] Y. Freund and R.E. Schapire, "Experiments with a new boosting algorithm," Proceedings of the 13th International Conference on Machine Learning, pp. 148–156, 1996.

[14] Y. Freund and R.E. Schapire, "A decision-theoretic generalization of on-line learning and an application to boosting," Journal of Computer and System Sciences, vol. 55, pp. 119–139, 1997.

[15] K. Cherkauer, "Human expert-level performance on a scientific image analysis task by a system using combined artificial neural networks," Working Notes of the AAAI Workshop on Integrating Multiple Learned Models, pp. 15–21, 1996.

[16] R. Ranawana und V. Palade, "A neural network based multi-classifier system for gene identification in DNA sequences," Neural Computing and Applications, vol. 14, pp. 122–131, 2005.

[17] T.G. Dietterich and G. Bakiri, "Solving multiclass learning problems via error-correcting output codes," Journal of Artificial Intelligence Research, vol. 2, pp. 263–286, 1995.

[18] Z. Ghahramani and H. Kim, "Bayesian Classifier Combination," Gatsby Technical Report, 2003.

[19] D.H. Wolpert, "Stacked generalization," Neural Networks, vol. 5, pp. 241–259, 1992.

[20] S. Dreiseitl and L. Ohno-Machado, "Logistic regression and artificial neural network classification models: a methodology review," Journal of Biomedical Informatics, vol. 35, pp. 352–359, 2002.

[21] F. Harrell, Regression Modeling Strategies: With Applications to Linear Models, Logistic Regression, and Survival Analysis, Springer, 2001.

[22] D. Hosmer and S. Lemeshow, Applied Logistic Regression, John Wiley & Sons, 2nd Edition, 2000.

[23] T. Hastie, R. Tibshirani, and J. Friedman, The Elements of Statistical Learning: Data Mining, Inference, and Prediction, Springer, New York, 2001.

[24] S. Dreiseitl, L. Ohno-Machado, S. Vinterbo, H. Billhardt, and M. Binder, "A comparison of machine learning methods for the diagnosis of pigmented skin lesions," Journal of Biomedical Informatics, vol. 34, pp. 28–36, 2001.

[25] T. Hastie and R. Tibshirani, "Discriminant adaptive nearest neighbor classification," IEEE Transactions on Pattern Analysis and Machine Intelligence, vol. 18, pp. 607–616, 1996.

15

The Use of Stochastic Processes in Bridge Lifetime Assessment

Khalid Aboura, Bijan Samali, Keith Crews and Jianchun Li

University of Technology, Broadway, Australia

In this chapter we introduce an approach for modeling the structural deterioration of components of bridges for maintenance optimization purposes. The Markov chain model can be found in the maintenance and repair problems since the early 60's, is introduced to the maintenance of road infrastructure in the 1980's, and is made to drive the current bridge maintenance optimization systems. While this model results into solvable programming problems and provides a solution, there are a number of criticisms associated with it. We highlight the shortfalls of the Markov model for bridge lifetime assessment and promote the use of stochastic processes.

15.1   Bridge Maintenance Optimization

Bridge maintenance optimization was applied these past decades due to the large costs associated with the management of networks of aging structures. In the United States, more than 70% of the bridges were built prior to 1935, and a large percentage of the United Kingdom's current bridge stock was built between the late 1950s and early 1970s. In the state of New South Wales, Australia, around 70% of the operating bridges were built before 1985, with a significant proportion before the 1940's. With the near completion of most of the road networks and the aging of bridges, the emphasis shifted to the maintenance and rehabilitation of the existing infrastructure. A concerted effort was made in the 1970's after the collapse of several bridges in the United States in the late 60's [1]. A number of mandates by governments introduced standards and computerized maintenance optimization approaches. These software tools, known as Bridge Management Systems (BMS), consist of formal procedures and methods for gathering and analyzing bridge condition data. The purpose of a BMS is to predict conditions for bridge stocks and estimate maintenance funding. Pontis [2], one of the most widely used systems, was designed and developed at the request of the US Federal Highway Administration. A similar BMS is the BRIDGIT bridge management system [3]. A bridge in a BMS is represented by structural elements defined as a set of common bridge components. For each bridge, the conditions of these elements are assessed visually during periodic inspections, and reported into the BMS. The condition data at each inspection consist of the total quantity of the element being divided through a number of condition states; from the 'as good as new' condition to the most severe state of deterioration. This is a common practice for representing infrastructure condition data, starting with the discrete condition rating scale from 0 to 9 adopted by the U.S. Federal Highway Administration [4], and followed by the Pontis condition rating scale, and that of most other bridge inspection procedures. This process simplifies the inspections, but more importantly the modeling of the element/bridge/network condition and the maintenance optimization. The condition assessment and prediction is done by the application of a Markov chain model. The Markov chain model can be found in the maintenance and repair problems since the early 60's [5]. It is introduced to the maintenance of road infrastructure by Golabi et al. (1982) [6], and is made to drive the Pontis bridge management system. In the Markov model, the condition of a bridge element takes discrete states and the transitions from one state to the other are modeled with a Markov chain. While this approach has become standard, there have been a number of criticisms associated with it. In this chapter, we discuss the fundamentals and argue for the applicability of the gamma process and other stochastic processes for modeling structural deterioration.

15.2   The Markov Chain Model

Markov model formulations are appealing to manage infrastructure because they provide a framework that accounts for the uncertainty and the optimal policies can be obtained by solving simple programming problems. A number of criticism points have been made against the usefulness of the model [1]; (i) bridge element performance is not addressed from a reliability viewpoint,

(ii) the Markovian assumption does not take into account the history of the bridge deterioration, and (iii) bridge system performance is not generally addressed. Among these limitations, a most important one is the inability of the Markov model, by assumption, to capture the time effect of deterioration. It is clear to many practitioners that the deterioration rate tends generally to increase with time, and amount of deterioration. This observation conflicts with the stationarity assumption of the Markov model where the transition probabilities do not change with time. In the context of the estimation of the Markov model transition probabilities, Madanat et al. (1995) [7] make a number of observations. They point to the fact that the methods used in estimating these probabilities are ad-hoc and suffer from important methodological limitations; (i) the change in condition from one inspection to the next is not modeled explicitly, failing to capture the structure of the deterioration process, (ii) consequently, the model fails to capture the inherent non-stationarity of the deterioration process, and (iii) the approach does not recognize the latent nature of deterioration. Since deterioration is an unobservable process, it is not the state of the observable condition that should be modeled, but rather the process that generates these conditions.

Transition Probabilities

A practical difficulty with the estimation of the transition probabilities is the lack of data for some condition states. The severe conditions of elements of a bridge are rarely observed, due to the maintenance effort. Only the first states transition probabilities are estimated properly. Figure 15.1 shows the frequency of units of different elements in the possible condition states, when inspected. An element on a bridge is measured in a number of unit, either in square meters (m^2) if it is a surface, in meters (m) for some elements such as railing and joints, or units (ea) for timber elements. The examples of Figure 15.1 are taken from a network of bridges in the study we mention in the next section. They show that most of the element units are in state 2 when they are not in the 'as good as new' or 'full condition' state 1. While it depends on the element and on budget, many elements are brought back to the full condition state from state 2. This implies that little data are available about transitions to more severe conditions. Pontis supplements the data with expert knowledge at the beginning of system implementation due to the scarcity of data at that stage. As more data becomes available, the probabilities are to be updated using a Bayesian method. This is a major shortfall, as many managers

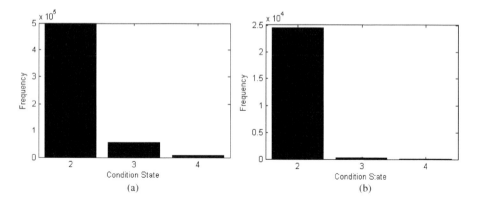

Fig. 15.1 Condition state frequency for a) Element example 1, b) Element example 2.

are reticent about the use of subjective assessment, particularly if there is little chance that enough data will appear to adjust the subjective input.

Stationarity Assumption

The theoretical weakness of the Markov model is the assumption of stationarity. The model assumes that the time of transition from one state to another is distributed exponentially. This assumption allows the application of the Markov model by which an old unit of element and a newer one are equally likely to move to the next condition [2]. This is a rigid structure imposed on the deterioration model. This weakness is acknowledged in most studies, and yet the Markov model has been adopted from the beginning and remains in most bridge management systems. It provides a solution for the estimation of future conditions and prescribes how to invest maintenance funding. A predictive solution relies on the validity of the underlying deterioration model. Scherer and Glagola (1994) [8], Wirahadikusumah et al. (2001) [9] and Morcous (2006) [10], among others have studied the validity of the state independence assumption of the Markov model. This assumption assumes that the future condition of a bridge element depends only on its present condition and not on its past condition, implying that bridge deterioration is a stationary process. While the simplifying assumption holds in some studies, it is not substantiated with a lot of evidence, the available condition data not providing adequate information about all possible condition transitions.

15.3 The RTA Level-2 Inspection Data

A study is conducted for the Roads and Traffic Authority of the state of New South Wales. The goal in the case study is to develop deterioration models for the assessment of future conditions of elements/bridges/networks of bridges. The data provided consist of inspection records conducted periodically on all bridges in Australia's largest state. The inspections are visual for most elements and the condition states are discrete. An element consists of a number of quantity units. Each unit is judged to be in one of n states, n = 3, 4 or 5, depending on the element type. Condition state 1 represents the 'as good as new' condition, no-deterioration state, while condition states 2...n mark increasing levels of deterioration. There are 4,945 structures and 66 elements considered in the study, for which over 230,000 inspection records exist. The records go back 15 years, and are up to the present time. Among the recorded entries is q, the total quantity of the element, q_1 the quantity in condition state 1 on the day of that inspection for that particular element, q_2 the quantity in state 2, to q_n the quantity in condition state n. The inspected quantities for each element are measured either in square meters (m^2) if it is a surface, in meters (m) for some elements such as railing and joints, or units (ea) for timber elements. The structures have on average about 10 elements, with more than 30 in some cases. Each element was inspected approximately every two years. The elements are concrete elements, steel elements with lead based paint, steel elements with other protective treatment, timber elements, joints, bearings, railings and others. Some of the more common types are concrete pre-tensioned girder, concrete reinforced pre-stressed pile, concrete deck slab, concrete culverts, steel rolled beams/I girders with lead based paint protective coating, timber beam/cross girder, pourable/cork joint seal, elastomeric bearing, metal bridge railing and masonry/brick/reinforced earth.

Condition Index

Deterioration in elements of bridges is best measured as a continuous variable representing for example the percentage of degradation. Given the large surfaces and number of items to be inspected, it is hard to measure visually percentages. The condition data at each inspection therefore consists of the total quantity q of the element being divided through judgment into n states. $q = \sum_{i=1}^{n} q_i$ is satisfied in all inspection records. The percentage of undamaged quantity is $q_1/q \times 100\%$. The corresponding proportion of deteriorated element is $q_d = \sum_{i=2}^{n} q_i / q$, $q_d = 1 - q_1 / q$. If one was to ignore the degree of deterioration and consider only the proportion of damaged quantity, then $C = q_1 / q \times 100$ provides a first level of information on the condition of the element. In our study, the condition data $q_1 \ldots q_n$ of an element is converted to a univariate measure C, using the notion of 'Condition Index'. A number of condition indices were formulated. These indices can be related to the California bridge health index [11], a ranking system that takes values in [0,100]. The California Department of Transportation was involved in the development and implementation of Pontis. A condition index has two functions; (i) its use in a cost/benefit analysis where the condition history of a structure can be estimated with the inspection data of its elements, through the use of a weighted sum of the conditions of the elements, and (ii) its use in the study of deterioration by modeling the univariate measure in time.

15.4   Estimating the Deterioration Rate

In turning the vector of quantities $q_1 \ldots q_n$ into a single value $C_{elem}(t)$ where t is time, the dimensionality of the problem is reduced and one can apply mathematically tractable models. This univariate quantity can be studied over time using a stochastic process. The element condition is a value between 0 and 100%, 100% being the 'as good as new' condition (no deterioration) state. An element can be in one of 11 states at an inspection time. The 11 states are mutually exclusive and exhaustive events, defined using two consecutive element conditions. Some of these states are (S_1) the element condition C is 100% and 100% at the previous inspection, (S_2) the element condition C is less than 100%, but positive, while 100% at the previous inspection, (S_3) the element condition C is 0% while 100% at the previous inspection, (S_7) the element condition C is less than 100%, but positive, and less than at the previous inspection where it was less than 100%, but positive, and (S_8) the element condition C is 0%, while it was less than 100%, but positive at the previous inspection. These states are of relevance when analyzing the deterioration. The others states are due to maintenance, when a renovation or repair improves the condition of the element or brings it back to 'as good as new'.

Deterioration Measure

The deterioration is defined as $Z = 100 - C$, $[C \equiv C_{elem}(t)]$. Z_1, Z_2, Z_3, Z_7 and Z_8 are the deterioration corresponding to states S_1, S_2, S_3, S_7 and S_8. These variables are of relevance when studying the deterioration process, along with their inter-inspection times. The time at which a deterioration increase, dZ in a time interval dt, occurs is important. The pairs (dt, dZ), for $dZ > 0$, and the time at which dt starts are information used to estimate deterioration.

Time Pattern

In order to study the distribution of the deterioration variables Z_2 and dZ_7 (the increase associated with Z_7), we need data that occur at the same time. In practice, inspections are not always strictly periodic. However, the inter-inspection times can be adjusted so that they are grouped together, making it possible to study the probabilistic behavior of deterioration as a random variable at different times. The adjustment was made through rounding off the inspection times. The pattern showed that the inspection times were meant to be 2 years apart. Figure 15.2 shows the conditions $C = 100 - Z_2$ plotted in dots, of some concrete element type chosen as an example, as they occur in time from a renewal, on the structures the element belongs to. The stars are the adjusted conditions.

Judgment was used in eliminating data for inspection time intervals that exceed 4 years. Beyond 4 years, the data start thinning. In the case of this

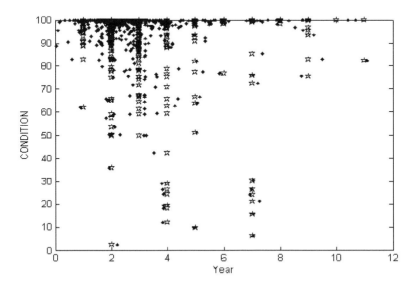

Fig. 15.2 Condition data of a concrete element type.

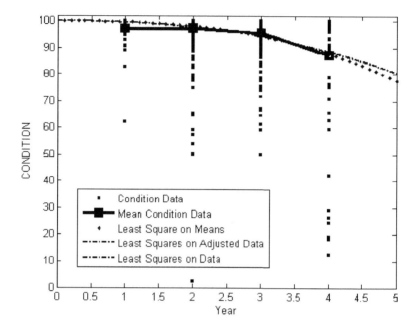

Fig. 15.3 Condition of a concrete element.

element, from the original 913 conditions, $C = 100 - Z_2$, 836 data points adjusted in time are plotted in Figure 15.3. Next, we used a least squares approach in fitting the data to a nonlinear deterioration curve $100 - \mu t^q$, with parameters μ and q. Figure 15.3 displays the least squares results using (i) the condition means at times 1,2,3 and 4 years, (ii) all the adjusted data, and (iii) the actual data. q was found to be 2.7, 2.4 and 2.4 respectively, and $\mu = 0.29$, 0.404, 0.408, showing that the adjustment to the data was minimal. The time pattern of the deterioration of the element fits the behavior of a stochastic process with mean function μt^q.

Probability Distribution Fit

The distributions of Z_2 at different points in time characterize the stochastic process. In the case of most of our elements, the probability distribution showed a good fit to the gamma distribution (Figure 15.4). This agrees with the properties of the gamma deterioration process, where the distributions of incremental deteriorations are gamma distributed [12]. This was observed for

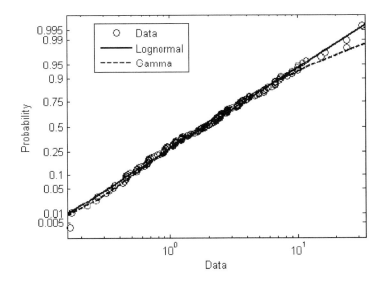

Fig. 15.4 Gamma and Lognormal probability distribution fits.

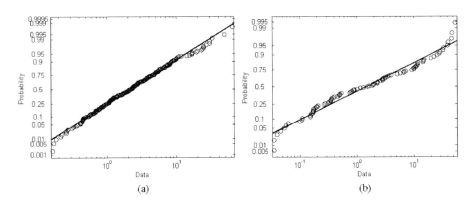

Fig. 15.5 Lognormal probability fit for a) Concrete element, b) Steel element.

the concrete and steel elements, as well as railing and joints and some timber elements. The lognormal was also found to be a good fit (Figure 15.5).

In the gamma process, not only is the distribution a gamma distribution when measured from the start of time, but any incremental distribution is also a gamma distribution. That is, if taking an interval in time and measuring the difference, increase, in the quantity of interest, the distribution of such an increase is also a gamma distribution. In addition, the increases in

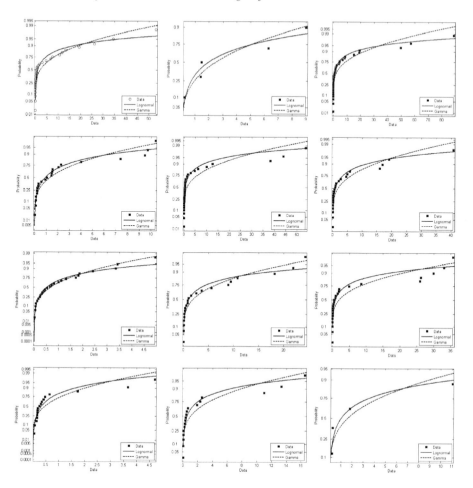

Fig. 15.6 Gamma and Lognormal probability fit for deterioration increase.

non-overlapping time intervals are independents. This makes for a number of assumptions to be checked. Figure 15.6 shows the distribution fit to the gamma and lognormal probability distributions for the increase in deterioration dZ_7, in a time interval $dt = 2$ years, when recorded at $t + dt$ years after a renewal. The plots in Figure 15.6 are for $t = 0, 1, 2, \ldots, 11$ left to right, top to bottom.

This example is taken for a concrete element common on a number of structures so that the data on dZ_7 allow a conclusive distributional fit. The good fit was also observed within classes of structures, following stratification of the data. We observed that whenever dZ_7 data are abundant enough, the stochastic

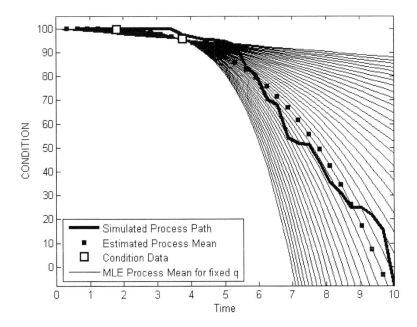

Fig. 15.7 Gamma process estimation.

process assumptions can be accepted. The data of the case study show the behavior of a stochastic process with selected probability distributions.

Gamma Process Application

For the purpose of building a statistical model, these observations promote the use of the gamma process with mean deterioration function μt^q that includes the stationarity case of $q = 1$. The gamma process is defined by van Noortwijk (2009) [12] in the context of the deterioration of structures. To illustrate the estimation procedure, we apply it to a concrete element and observe a rate $\hat{q} = 3.3$ (Figure 15.7). Two data points are used; (15.81, 99.69) and (17.75, 95.46) with the renewal at year 13.92 since the start of the database. This example was chosen for illustrative purposes. In many cases, the bridge elements showed lower rates.

15.5 The Gamma Process

The gamma process can capture the temporal variability of degradation. The argument is made in a series of papers by Pandey and van Noortwijk

(2004) [13], van Noortwijk et al. (2005) [14] and Pandey et al. (2007) [15]. The gamma process can be found in its modern application to structures in the late 90's by van Noortwijk (1998) [16] and van Noortwijk and Klatter (1999) [17]. The idea of the use of a gamma process can be found earlier in the Netherlands where generalized gamma processes were used to model decision problems for optimizing maintenance of the sea-bed protection of the Eastern-Scheldt barrier, berm breakwaters, and dykes [18]. Empirical studies showed that the expected deterioration in some cases followed the power law at^b, where t is the time. This function of time is incorporated into the gamma process and used to model structural deterioration. The advantage of the gamma process is recognized and applied in many structural studies [19, 20, 21]. van Noortwijk (2009) [12] provides a comprehensive overview of the use of the gamma process in the maintenance of structures.

Definition

In the context of structural deterioration, the gamma process is defined as follows: Let $v(t)$ be a non-decreasing, right continuous, real-valued function for $t > 0$, with $v(0) = 0$. The gamma process with shape function $v(t) > 0$ and scale parameter $u > 0$ is a continuous-time stochastic process $Z(t), t > 0$ with the following properties;

    (i) $Z(0) = 0$ with probability 1,
    (ii) $Z(\tau) - Z(t) \sim G(v(\tau) - v(t), u)$ and
    (iii) $Z(t)$ has independent increments,

where $G(z|v, u) = u^v z^{v-1} e^{-uz} / \Gamma(v)$ is the gamma probability density function defined for $z > 0$. The process can be parameterized. Letting $v(t) = \mu^2 t^q / \sigma^2$ and $u = \mu/\sigma^2$, the mean and variance of the deterioration $Z(t)$ are:

$$E(Z(t)) = \mu t^q \tag{15.1}$$

$$V(Z(t)) = \sigma^2 t^q \tag{15.2}$$

Given a set of observations of the deterioration process $Z(t)$, $\{z_i\}_{i=1}^n$ for times $\{t_i\}_{i=1}^n$, the maximization of the likelihood function provides estimates of the three parameters (μ, σ, q). This involves the search, q fixed, for the zero of a function, where $\hat{\sigma}$ is solution of

$$\sum_{i=1}^n w_i \left\{ \psi\left(\frac{\hat{\mu}^2}{\sigma^2} w_i\right) - \log\delta_i \right\} = t_n^q \log\left(\frac{x_n}{t_n^q \sigma^2}\right) \tag{15.3}$$

where $\delta_i = z_i - z_{i-1}$ and $w_i = t_i^q - t_{i-1}^q$, $i = 1, \ldots, n$, $z_0 = 0$ with $t_0 = 0$, and $\hat{\mu} = z_n / t_n^q \cdot \psi$ is the Digamma function.

Simulation and Estimation

We experimented with the process using simulation and estimating the parameters with the maximum likelihood approach. With few data points, the model captures the deterioration process efficiently. The lower the data points are along the curve, the better the estimation. To illustrate, we assume that three data points are observed; $(y_1, c_1) = (2, 99.5)$, $(y_2, c_2) = (4, 99.4)$ and $(y_3, c_3) = (6, 95.2)$. This means that some years after the start of the database, at a renewal point, the element deterioration is $z_0 = 0$, time is reset to $t_0 = 0$ and the condition C_0 is set to $100(\%)$. At the next inspection, 2 years later, (y_1, c_1) is recorded leading to $(t_1, z_1) = (2, 100 - c_1) = (2, 0.5)$. Similarly, (t_2, z_2) and (t_3, z_3) obtain. The first deterioration is of 0.5% magnitude, then 0.1%. Then a further 4.2% of the element deteriorates. Varying the deterioration rate of the process, $q = 1, \ldots, 5\mu$ and σ are estimated using the maximum likelihood method for each q value. The resulting process mean functions $100 - \hat{\mu}t^q$, for corresponding q, are plotted in Figure 15.8. $\hat{\mu}$ is the maximum likelihood estimate of μ for given q. Applying the maximum likelihood principle again, the deterioration rate is estimated to be $\hat{q} = 2$, for corresponding $\hat{\mu} = 0.13$ and $\hat{\sigma} = 0.48$. The estimated condition process mean function $100 - \hat{\mu}t^{\hat{q}}$ is shown as a dotted line in Figure 15.8.

Using the estimated parameters $(\hat{\mu}, \hat{\sigma}, \hat{q})$, a condition path is simulated for illustrative purposes and shown in Figure 15.9.

Replicating 50 times, the condition data are shown in Figure 15.10. The data points are not connected with lines, to better illustrate the behavior of the stochastic process at a fixed point in time. For a given point in time, the corresponding data are gamma distributed.

Figure 15.11 shows the theoretical mean deterioration process of a gamma process, along with the ensuing condition mean process. The theoretical distributions of the accumulated deterioration (Z_7) and the increase deterioration (dZ_7) are schematized.

Accuracy of the Estimation Procedure

In experimenting with the gamma process, we observe that with few data points the model captures the deterioration process efficiently. The lower the data points are along the curve, the better the estimation. This is a significant

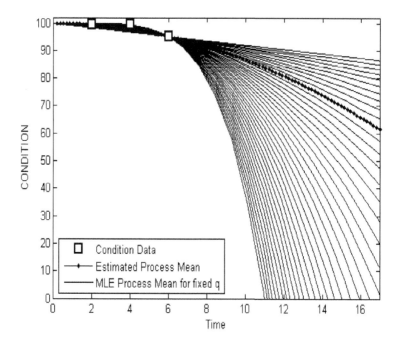

Fig. 15.8 Maximum likelihood estimation.

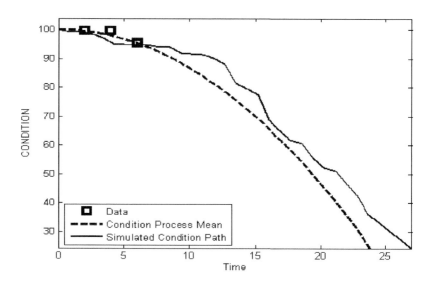

Fig. 15.9 Simulated condition path.

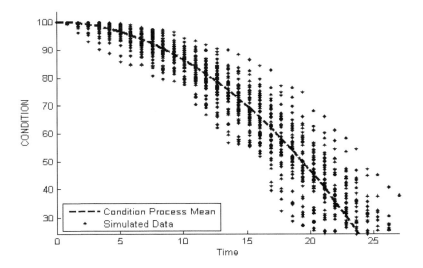

Fig. 15.10 Simulated condition paths for 50 replications.

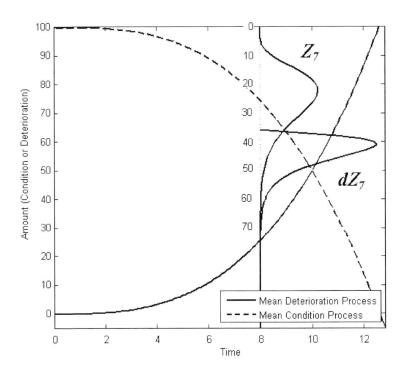

Fig. 15.11 Theoretical mean deterioration and condition processes.

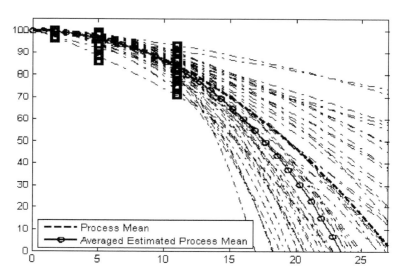

Fig. 15.12 Averaged estimated process means.

observation. The high deterioration points for an element of a bridge are not many and the few that are observed do not provide good estimates for the Markov transitions to high deterioration states. However, few of these data points provide a good estimation when using a gamma process model. To illustrate this point, we choose to simulate 50 condition paths using the same estimated parameters, whose theoretical process mean is shown by the dashed line in Figure 15.12. For each simulated path, three data point are taken, using the times $t_1 = 1.6875$, $t_2 = 5.0625$ and $t_3 = 10.9688$. For each path, the data may differ, and are plotted as small squares in the figure. The resulting MLE estimated condition process means are shown in Figure 15.12. Their average is plotted as a line with small circles.

It is clear that the estimation is more efficient at the top of the curve where the data lie. This can be verified theoretically for a fixed q, since the estimate of $\hat{\mu}$ depends on the last data point (t_n, z_n). However, in estimating the triplet (q, μ, σ), a closed form solution is not available. We observe empirically that the lower the condition data is on the curve, that is the higher the observed deterioration, the better the estimation of the condition (deterioration) process. Keeping $t_1 = 1.6875$ and $t_2 = 5.0625$ but varying t_3, we observe the average estimated process mean perform better as the third data point goes deeper in time, and hence in deterioration (see Figure 15.13). At $t_3 = 22.78$, the

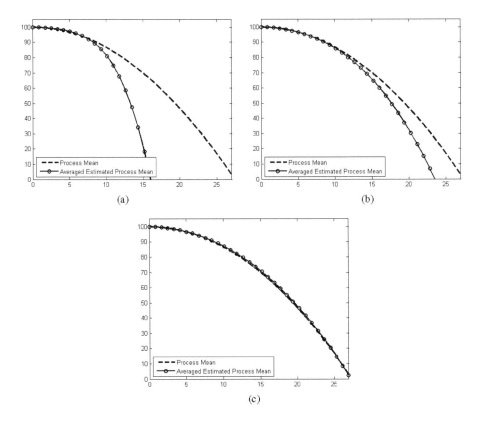

Fig. 15.13 Average estimated process mean for a) $t_3 = 6.75$, b) $t_3 = 10.96$, c) $t_3 = 22.78$.

deterioration is 69.2%, for a corresponding condition of 30.8%. This leads, on average, to a perfect estimation of the process mean curve.

This point is further illustrated in Figure 15.14 where we choose $t_3 = 24.4688$, for a corresponding 80% theoretical deterioration. Figure 15.14 shows the corresponding statistics of the 50 simulation-MLE results.

For an element whose deterioration behaves in time according to a gamma process, any information on the condition path leads to a better estimation. But the deeper that information is along the curve, the higher the observed deterioration, the better the theoretical process mean is estimated. The importance of such estimation is apparent in the execution of the final exercise, the maintenance optimization. In estimating at what time the condition of an element will reach a target level, and the ensuing economic consequences of

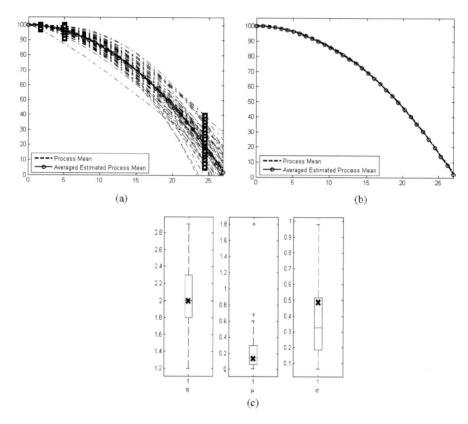

Fig. 15.14 Average estimated process mean with deep information a) Simulation, b) Averaged process, c) MLE statistics.

corresponding maintenance times, the accuracy is of significant importance. It is illustrated in Figure 15.15, where two different condition paths lead to two very different times, t_1^* and t_2^*, to reach the 20% deterioration target. There is approximately a 2 years difference between the two maintenance times, which can be of significant difference in terms of cost if applied to a large number of elements on different bridges or to expensive elements.

Aggregating Data

The exercise of estimating the process mean curve is at the heart of the maintenance optimization problem. While the decision model requires the declaration of costs, a non trivial part, estimating properly the condition curve is the essential part of the problem. The data for each individual bridge aren't enough to

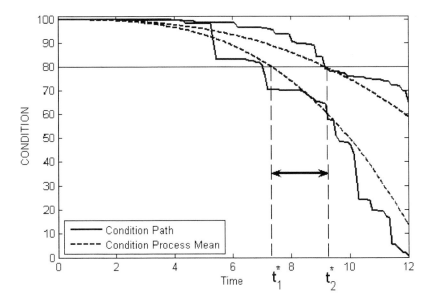

Fig. 15.15 Time to reach target level.

estimate the parameters μ, σ and q properly. Often, the path $\{z_i\}_{i=1}^n$ for times $\{t_i\}_{i=1}^n$ does not extend beyond n = 1 or 2, before the element is brought back to the full condition state $Z = 0$. One way around this problem is to aggregate the data by dividing the bridges into similarity classes. Bridges are grouped through the identification of major elements. Then within the classes of similar bridges, a second stratification occurs according to influencing factors; traffic load, age, region and environmental stress. The estimation of the parameters within the class structures results in an effective deterioration assessment and prediction.

The approach of aggregating data is the one taken in most scenarios. Regardless of the statistical model, be it the Markov chain or a stochastic process such as the gamma process, one is often forced to pull information about an element type from different bridges and estimate and predict the deterioration. But in the gamma process, the approach provides a more efficient estimation procedure. While few data points for high deterioration provide at best weak estimates for the transition probabilities to the corresponding states in the Markov model, in the gamma process, they help pin down the deterioration curve more accurately, as seen in the simulation analysis.

First Observation Data

In the special case where only the first deteriorations, observed at the first inspection after a renewal, are available, the solution is similar to the one for a single path seen above, but the search for $\hat{\sigma}$, for fixed q, involves the equation

$$\sum_{i=1}^{m} w_i \left\{ \psi \left(\frac{\hat{\mu}^2}{\sigma^2} w_i \right) - \log \delta_i \right\} = \left(\sum_{i=1}^{m} w_i \right) \log(\frac{\hat{\mu}}{\sigma^2}) \qquad (15.4)$$

where z_i is the deterioration of the ith element in the class of structures, $\delta_i = z_i$ and $w_i = t_i^q$, $i = 1, \ldots, m$ and $\hat{\mu} = \sum_{i=1}^{m} z_i / \sum_{i=1}^{m} t_i^q$. In general, due to the independent increments property and with the assumption of independence between elements on different structures, the likelihood function can be written and maximized to estimate the parameters μ, σ and q [21].

15.6 Lognormal Diffusion Process

Another candidate for the modeling of the deterioration is the lognormal diffusion process. Used in statistical modeling, it has been mainly confined to the analysis of economic data [22]. The lognormal diffusion process with exogenous factors is defined by the process $\{X(t), t \in (t_0, T, t_0 \in \Re\}$ with positive values and infinitesimal moments:

$$A_1(x, t) = x h(t) \qquad (15.5)$$
$$A_2(x, t) = \sigma^2 x^2 \qquad (15.6)$$

where $h(t) = \beta_0 + \sum_{j=1}^{q} \beta_j F_j(t)$, $\beta_j \in \Re$, $\sigma > 0$ and F_j are time-continuous functions in $(t_0, T$ $j = 1, \ldots, q$. The transition probability density function of this process is a lognormal distribution and the trend is given by the expression

$$E(X(t)) = E(X(t_0)) exp \left(\int_{t_0}^{t} h(\tau) d\tau \right) \qquad (15.7)$$

The estimation of the parameters $(\beta_0, \ldots, \beta_q, \sigma)$ is conducted through the maximization of likelihood as for the Gamma process. Replacing the process mean μt^q of the gamma process with a trend $e^{\lambda t^q} - 1$ in the lognormal diffusion process can capture the deterioration process.

15.7 Summary of Chapter

Structural deterioration assessment and condition prediction of bridges is a subject of interest that has far reaching consequences, both in terms of public safety and budgeting. This chapter discusses the theoretical foundations of the model most used for assessing structural deterioration in bridge elements. A modern view is proposed using the gamma process that has proved successful in structural deterioration assessment. Experimentation with the process is shown using simulation and estimation of the parameters of the model. With few data points, the model captures the deterioration process efficiently. In the case study, a probability distribution fit is observed that supports the use of the gamma process as well as leads to the consideration of other stochastic deterioration processes.

References

[1] D.M. Frangopol, J.S. Kong and E.S. Gharaibeh, "Reliability-based life-cycle management of highway bridges", Journal of Computing in Civil Engineering, vol. 15, no. 1, pp. 27–34, Jan. 2001.

[2] K. Golabi and R. Shepard, "Pontis: A system for maintenance optimization and improvement of US bridge network", Interfaces, vol. 27, no. 1, pp. 71–88, Jan.–Feb. 1997.

[3] H. Hawk and E.P. Small, "The BRIDGIT bridge management system", Structural Engineering International, vol. 8, no. 4, pp. 309–314, Nov. 1998.

[4] Federal Highway Administration, Recording and coding guide for structure inventory and appraisal of the nation's bridges, U.S. Department of Transportation, Washington, D.C., U.S.A, 1979.

[5] M. Klein, "Inspection-maintenance-replacement schedules under Markovian deterioration", Management Science, vol. 9, no. 1, pp. 25–32, Oct. 1962.

[6] K. Golabi, R.B. Kulkarni and G.B. Way, "A Statewide Pavement Management System", Interfaces, vol. 12, no. 6, pp. 5–21, Dec. 1982.

[7] S. Madanat, Mishalani, R. and W.H. Wan Ibrahim, "Estimation of infrastructure transition probabilities from condition rating data", Journal of Infrastructure Systems, vol. 1, no. 2, pp. 120–125, June 1995.

[8] W.T Scherer and D.M. Glagola, "Markovian models for bridge maintenance management", Journal of Transportation Engineering, vol. 120, no. 1, pp. 37–51, 1994.

[9] R.Wirahadikusumah, D. Abraham and T. Iseley, "Challenging issues in modeling deterioration of combined sewers", Journal of Infrastructure Systems, vol. 7, no. 2, pp. 77–84, June 2001.

[10] G. Morcous, "Performance Prediction of Bridge Deck Systems Using Markov Chains", Journal of Performance of Constructed Facilities, vol. 20, no. 2, pp. 146–155, May 2006.

[11] R.W. Shepard and M.B. Johnson, California Bridge Health Index, IBMC-005, California Department of Transportation, International Bridge Management Conference, Denver, Colorado, Preprints, Volume II, K1, 1999.

[12] J.M. van Noortwijk, "A survey of the application of gamma processes in maintenance", Reliability Engineering and System Safety, vol. 94, pp. 2–21, 2009. Available online March 2007.

[13] M.D. Pandey and J.M. van Noortwijk, "Gamma process model for time-dependent structural reliability analysis", In Bridge Maintenance, Safety, Management and Cost, E. Watanabe, D.M. Frangopol, and T. Utsonomiya, editors, Proceedings of the Second International Conference on Bridge Maintenance, Safety and Management (IABMAS), Kyoto, Japan, 18–22 October 2004. London: Taylor & Francis Group.

[14] J.M. van Noortwijk, M.J. Kallen and M.D. Pandey, "Gamma processes for time-dependent reliability of structures", In Advances in Safety and Reliability, K. Kolowrocki, editor, Proceedings of ESREL 2005 — European Safety and Reliability Conference 2005, Tri City (Gdynia-Sopot-Gdansk), Poland, 27-30 June 2005, pp. 1457–1464. London: Taylor & Francis Group.

[15] M.D. Pandey, X.-X. Yuan and J.M. van Noortwijk, "The influence of temporal uncertainty of deterioration on life-cycle management of structures", Structure and Infrastructure Engineering, pp. 1–12, April 2007.

[16] J.M. van Noortwijk, "Optimal Replacement Decisions for Structures under Stochastic Deterioration", Proceedings of the Eighth IFIP WG 7.5 Working Conference on Reliability and Optimization of Structural Systems, Krakow, Poland, 1998, pp. 273–280, Ann Arbor: University of Michigan.

[17] J.M. van Noortwijk and H.E. Klatter, " Optimal inspection decisions for the block mats of the Eastern-Scheldt barrier", Reliability Engineering and System Safety, vol. 65, pp. 203–211, 1999.

[18] J.M. van Noortwijk and P.H.A.J.M. van Gelder, "Optimal maintenance decisions for berm breakwaters", Structural Safety, vol. 18, no. 4, pp. 293–309, 1996.

[19] F.A. Buijs, J.W. Hall, J.M. van Noortwijk and P.B. Sayers, "Time-dependent reliability analysis of flood defences using gamma processes", In Safety and Reliability of Engineering Systems and Structures, Augusti G, Schueller GI, Ciampoli M, editors., pp. 2209–2216. Rotterdam: Millpress; 2005.

[20] J.M. van Noortwijk, J.A.M. van der Weide, M.J. Kallen, M.J. and M.D Pandey, "Gamma processes and peaks-over-threshold distributions for time-dependent reliability", Reliability Engineering and System Safety, vol. 92, pp. 1651–1658, 2007. Available online December 2006.

[21] R.P. Nicolai, R. Dekker and J.M. van Noortwijk, "A comparison of models for measurable deterioration: An application to coatings on steel structures", Reliability Engineering and System Safety, vol. 92, pp. 1635–1650, 2007. Available online November 2006.

[22] R. Gutierrez, P. Roman, and F. Torres, "Inference on some parametric functions in the univariate lognormal diffusion process with exogenous factors", Test, vol. 10, no. 2, pp. 357–373, 2001.

Index

RIVER PUBLISHERS SERIES IN INFORMATION SCIENCE AND TECHNOLOGY